MBI Publishing Company

POWERPRO SERIES

RACER'S ENCYCLOPEDIA OF
METALS, FIBERS
& MATERIALS

Forbes Aird

First published in 1994 by MBI Publishing Company, PO Box 1, 729 Prospect Avenue, Osceola, WI 54020-0001 USA.

MBI Publishing Company books are also available at discounts in bulk quantity for industrial or sales-promotional use. For details write to Special Sales Manager at Motorbooks International Wholesalers & Distributors, 729 Prospect Avenue, PO Box 1, Osceola, WI 54020-0001 USA.

Library of Congress Cataloging-in-Publication Data
Aird, Forbes.
    Racer's encyclopedia of metals, fibers & materials/Forbes Aird.
        p. cm.—(MBI Publishing Company powerpro series)
    Includes index.
    ISBN 0-87938-916-8
    1. Automobiles—Materials. I. Title. II. Series.
TL154.A517   1994   94-26704
629.23'2—dc20

On the front cover: An assortment of automotive and motorcycle parts and components made of materials such as aluminum, carbon fiber, rubber, and Childs & Albert's Dura-Moly. Connecting rods and piston rings courtesy of Childs & Albert Performance Products; Ducati 900SS carbon fiber fender courtesy of John Luthard; various other parts courtesy of Jim Kanan, Stillwater Motor Company. *Craig Lassig*

Printed in the United States of America

Contents

Dedication

I dedicate this book to the memory of Harvey Long, who taught me the rudiments of welding, blacksmithing, and heat treatment, and so much more, and how to swear colorfully when you grab hold of the hot end. Thanks old pal.

Acknowledgments

Please refer to the Selected Bibliography/Suggestions for Further Reading listings that start on page 124.

Introduction

Everything (at least, every tangible artifact) is made out of something. We call the different kinds of stuff that things are made from "materials," and there are a lot of them. For reasons that will become apparent, comparatively few are used in high-performance vehicles such as race cars and aircraft. This book represents an attempt to survey some of the more significant of those. While the contents will probably be of greatest interest to people who design, build, modify, or repair high-performance vehicles, it is hoped that others not actively involved—enthusiastic students, race fans, and observers—might also find something of value here.

As in any other field, there is a bit of specialized terminology to learn and understand. The first section of the book, then, explains the basic concepts and vocabulary of materials science, and tries to illustrate how the properties of a structure interact with the properties of the material it is made from. Later sections provide descriptions of individual materials, grouped for convenience into families. As it happens, the order in which the families are presented corresponds (roughly and in general) to the order in which they came into use—thus iron and steel precede light metals, and metals appear before composites. Notably, as we move from "older" to "newer," we also tend to move away from "found" materials—those which nature provides, if only in the form of ores—toward man-made materials, such as high-performance composites.

If nothing else, it is hoped that the reader will become aware that there is no one all-around "best" material. A decision about which material to use for any specific job should not amount to finding a way to use some new "trick stuff." The process of selection should proceed backwards—first describe the important characteristics required, then select the material that best meets those needs, within the budget available. As obvious as this seems, there have been many cases where a designer has become obsessed with one particular outstanding property of a material and applied it inappropriately. Welded aluminum space frames, titanium spindles, and ceramic valves are examples mentioned in these pages where the usual result is not improved performance but rather a load of trouble and grief.

Any such survey must necessarily be incomplete; certainly this one is. And while care has been taken to ensure accuracy, it is possible (indeed, most likely!) that there are errors as well as omissions, if only because the material presented here has been gathered from myriad diverse sources. Many of these are noted in the bibliography. Readers are invited to comment, criticize, and propose additions for a future edition. Contact me via the publisher.

Forbes Aird

PART 1

Chapter 1

The Properties of Materials

Stone Age; Bronze Age; Iron Age—the history of mankind is divided into periods named for materials. It is worth bearing in mind that these stages were not named by materials scientists or engineers, but by historians and other social scientists. The message is obvious: the materials we use—to obtain food, to build shelter, to make tools—are basic to the nature and quality of our lives.

As we have advanced from one era to the next, that is, from one material to the next, there is a general trend for the strength of the available materials to increase. Stronger material means not only that we can do some things we couldn't do before, it also means we can do the old things with less material. Now, apart from the labor required to hoist stones on top of each other, it is not immediately obvious that there is much of an advantage to using, say, less stone to make a building, or a bridge. But if you want a very tall building, or a bridge with a long span, or if you want an object to be portable—and especially if you want it to be mobile, its weight becomes important.

At this point it is worth pausing to point out that the idea of mobile objects has been around for a lot longer than steamships and railways and automobiles and aircraft. What stood in the way was not imagination or inventiveness—take a look at the work of Leonardo da Vinci—it was materials sufficiently light in relation to their strength and stiffness to make such self-propelled objects practical. (Actually, when we say "light" or "heavy" we really mean *density*—how much one cubic inch of the stuff weighs).

Now density is a pretty easy concept to grasp—you can judge it, roughly, by just grabbing hold of a chunk of the stuff and hefting it. Thus, wood is light; aluminum is heavier than wood, but lighter than steel; steel is heavy, and so is stone. And so on. Strength and stiffness (at least in the engineering sense) are not quite so straightforward, however.

All materials deform when they are loaded, and all materials break when the load becomes too large, but one of the attractions of metals— compared to other structural materials like wood, concrete, or fiberglass— is this ability to yield.

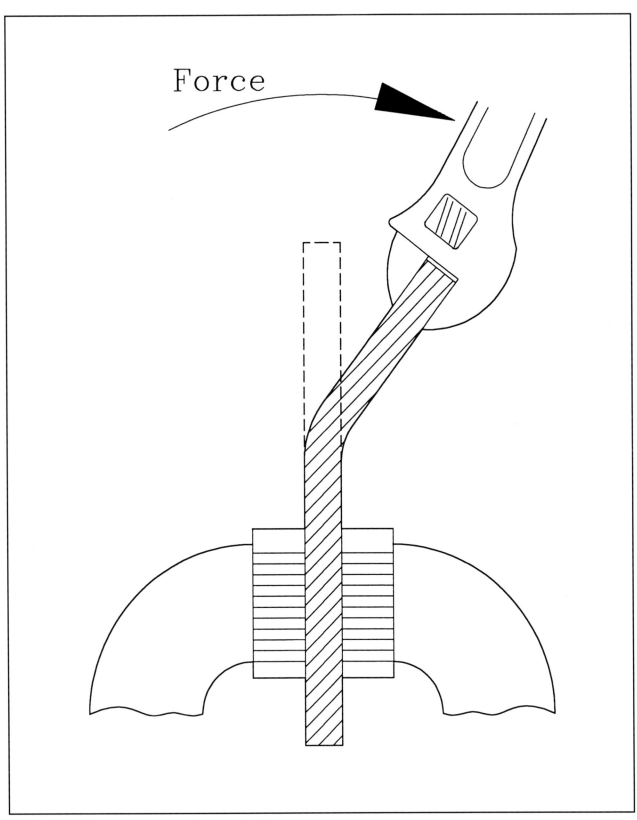

Force

All materials deform when loaded. If the applied force is
small, the distortion will usually take the form of "elastic
strain"...

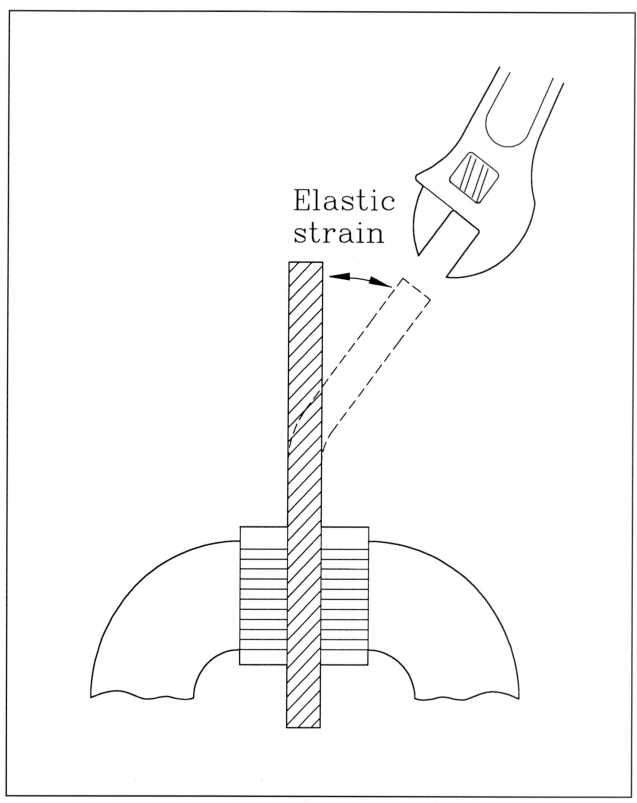

Elastic strain

...and the material will spring back to its original shape when the force is removed. If the force is large enough, damage will result. If the part is made from a brittle material, it will simply fracture. But if the material is a ductile metal...

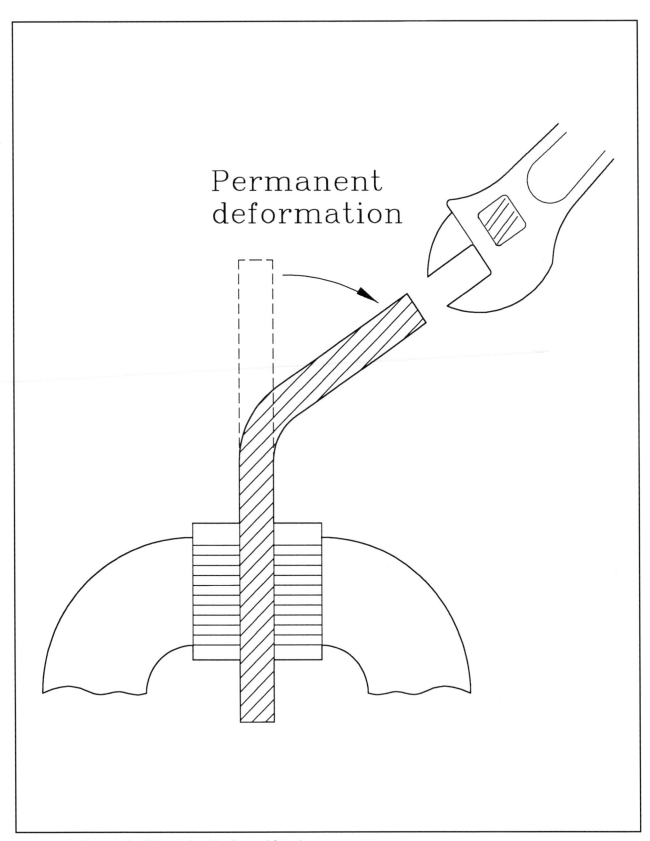

Permanent deformation

...the metal will yield—it will be noticeably distorted from its original shape, but remain in one piece.

Tensile Strength

Non-engineers are often puzzled by the way the strength of materials is expressed in engineering handbooks and suppliers' catalogs—the units are pounds-per-square-inch, which looks like a measure of *pressure* rather than of strength. To explain, imagine a square bar, measuring 1 inch (in) on each side. The cross section area of the bar, measured anywhere along its length, would obviously be 1 square inch (sq in).

Now, fasten one end of that bar to a (strong!) ceiling, and hang a 1,000 pound (lb) weight on the other end. That square inch of cross section now has 1,000lb of force passing through it, trying to pull each piece of bar apart from the adjacent section. Twice as much weight (or a bar with half the cross section area) would mean 2,000lb of force acting on each square inch of cross section. To avoid having to talk about both the load *and* the size of the bar, engineers use the term *stress* to describe the force-per-unit-of-cross-section-area. In our first case, the stress is 1,000lb per square inch—1,000psi. Simple enough?

Obviously, if you raise the stress to a high enough level, the bar will eventually break. The breaking point is called the *ultimate tensile strength* (UTS). If the 1sq in bar held 50,000lb but broke when more weight was added, the UTS of the material the bar is made out of is 50,000psi. A bar of 2sq in area would obviously carry twice as much load before breaking, but the stress would be the same 50,000psi.

Yielding

If the bar is made out of steel—or any other common structural metal, for that matter—an interesting and valuable effect occurs just before the breaking point is reached. If you carefully measured the length of the bar before and after you hung weights on it, you would find that the bar stretches slightly under moderate loads, then springs back to its initial length when the load is removed. But, as the load approaches the ultimate strength of the material, some of the stretch becomes permanent—the bar does not return to its original length when the load is removed. The stress level needed to permanently deform the material in this way establishes the *yield strength* of the material.

All materials deform when they are loaded, and all materials break when the load becomes too large, but one of the attractions of metals—

compared to other structural materials like wood, concrete, or fiberglass—is this ability to yield. When overloaded, they stretch and warp and generally behave like high-strength taffy, which is in many ways preferable to something of greater strength, but which lets go with a bang if you lean on it too hard.

From the fabricator's viewpoint, it is this property that makes possible bending, folding, pressing, and all other metal-shaping operations. From the designer's perspective, the range between yield strength and ultimate strength means that metals are generally forgiving materials, and the size of the spread between the early warning of yielding and total failure at the ultimate strength is a measure of the amount of "forgiveness" available.

Although a part that has yielded may retain significant strength, and might continue to do a useful job of load carrying, such a component has permanently changed its shape, and so must be considered to have failed. Therefore, the maximum load the material should ever experience in service, as opposed to shaping, is represented by the yield strength, not the ultimate strength. The additional margin available after yielding should be regarded as insurance.

Among the metals, titanium, super alloy steels, and some non-ferrous super alloys are about tied for first place in specific strength. . . . And because their yield and ultimate strengths are so close together, they are extremely sensitive to notches, holes, and other abrupt changes in cross section that create stress concentrations.

Shear, Compression and Bending

Shear

While some parts in race cars (and all other kinds of machinery) are stressed in tension, other parts bear different kinds of loads. In some cases, the load tries to slice the material, rather than pull it apart lengthwise. That's called a *shear* load. The examples most often quoted are flywheel bolts and chassis fasteners, but parts that are loaded in torsion, too, like coil and torsion bar springs and transmission shafts, are stressed in shear—at any point along the part, each section is trying to slide over the adjacent one.

For structural metals, the ultimate shear strength is about 60 percent of the ultimate ten-

The way a bending stress is resolved into tension, compression, and shear depends on the geometric arrangement of the material, so strength and stiffness in bending is more a property of a structure than of a material.

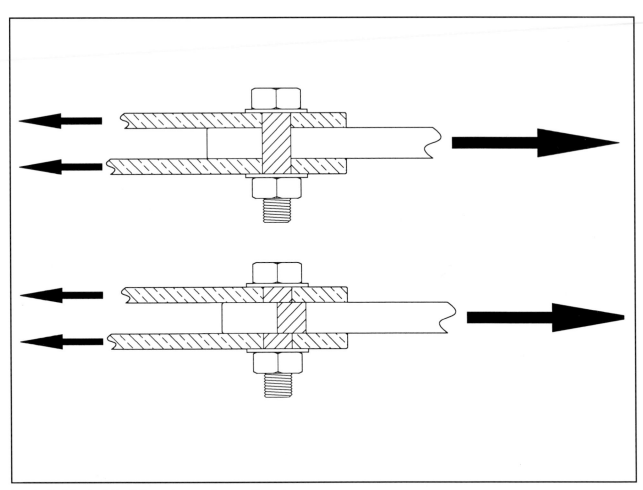

Loads that try to "slice" through a part are called shear loads. They can arise directly, as in this illustration, or they can result from torsion in a shaft or tube, or from edge-wise loads in a plate or sheet.

sile strength. (This figure is usually regarded as exactly correct for properly heat-treated steel; for other metals it is mighty close.) Likewise, there is a pretty regular relationship between the stiffness values in shear and tension (see "Stiffness," below)—the stiffness in shear for most metals is about 35 percent—38 percent of the stiffness in tension.

With other materials, the performance in shear cannot be predicted on the basis of tension properties. Not only do the actual values vary widely, they also depend strongly on the direction the shear forces run. Composites, for instance, show huge variations in both shear strength and stiffness according to whether the forces act across the fibers, or try to separate the fibers from each other. In a composite "laminate" (a piece consisting of multiple layers laid one over another), certain kinds of loads involve forces that work to slide one layer over another. The resistance to this "inter-laminar shear" depends almost entirely on the properties of the matrix in which the strong fibers are embedded (see chapter 9), and can reach very low values. And everyone understands that wood splits much more easily along the grain than across it.

Compression

If you push on a part, rather than pulling on it, the forces load the material in compression. If the component is relatively long and thin, it will buckle before the material will actually fail—you can't push on a rope, for instance. What's going on here obviously has more to do with the shape of the part than with the material it is made from—the strength and stiffness are properties of the structure, rather than of the material. (Such issues are dealt with in chapter 2.)

If the part is relatively short and squat, however, it may fail from pure compressive stress. Just how it fails will depend on whether its material is brittle or "ductile." If the material can yield, like common grades of steel or aluminum, it will squish like a stomped grape. (Remember leaving a penny on a railroad track, and what happened to it when the train went by? Pennies are made from copper, and copper is ductile—it yields). If the material is brittle, like cast iron or glass, there will be a noise like a cannon shot, and the piece will fracture—often along a line at 45 degrees to the applied force, although some materials will crumble to powder under the same circumstances.

Frames that lack stiffness cannot provide a stable platform for the suspension; gears that warp when heavily loaded lose their precise mesh, and so cost horsepower through extra friction; a springy crank throws off valve and ignition timing, and overloads the edges of engine bearings.

Many materials, including most structural metals, and some glass and carbon composites, have compression properties (strength and stiffness) similar to those in tension. But there are some exceptions. Cast iron, for instance, is much stronger in compression than in tension. On the other hand, aramid composites (see chapter 8), have similar tensile and compressive stiffness, but their compression strength is low, despite the fact that they are among the strongest of all materials in tension. Wood also generally shows substantially less strength and stiffness in compression than in tension. Magnesium, too, exhibits different properties according to the direction it is loaded. Its yield strength in compression is less than in tension by as much as 30 percent, depending on the alloy, although this is more generally true of wrought as opposed to cast "mag."

Bending

From the point of view of the atomic bonds that hold materials together, there are really only three kinds of loads—tension, compression, and shear. All loads can be represented as some combination of these three, just as you can produce any shade of paint by mixing the three primary colors. We have already suggested, for instance, that torsional loads, such as exist in shafts trans-

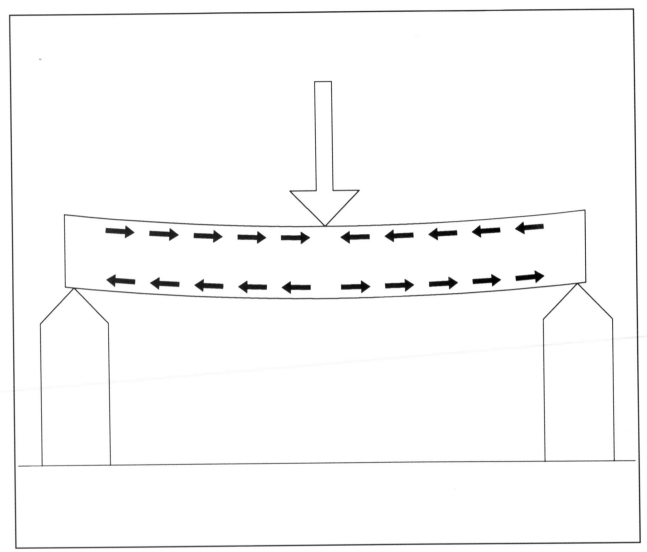

A bending or "beaming" force resolves into tension in one face of the beam and compression in the other.

mitting torque, actually stress the material in shear.

Bending—or "flexure"—is actually a combination of tension and compression, and usually involves shear, too. This isn't a case of sucking and blowing at the same time—a sample of material exposed to a bending force will have one surface in compression, while the other is in tension. If the material has similar strength and stiffness in both tension and compression, it will be a toss-up whether the tension side starts to rip asunder first, or the compression side starts to squish or wrinkle. If the material has properties favoring loads of one kind or the other, we can predict which side will pack up first. Concrete or cast

If the material has similar strength and stiffness in both tension and compression, it will be a toss-up whether the tension side starts to rip asunder first, or the compression side starts to squish or wrinkle.

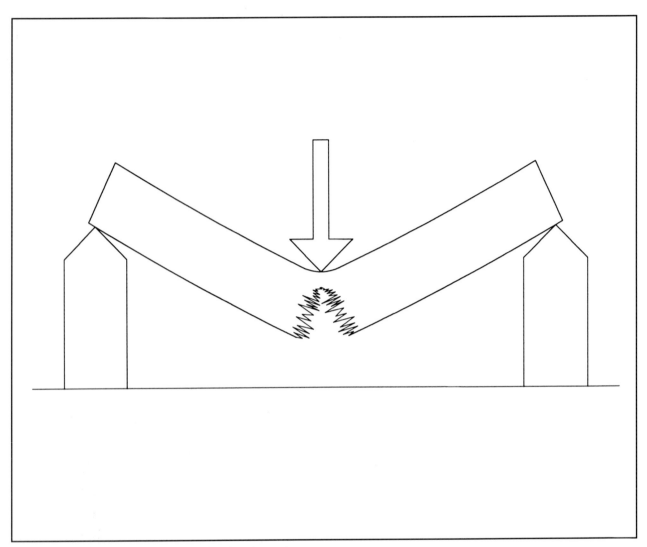

A material that is stronger in compression than in tension, like cast iron or concrete, will rupture on the tension side

iron will let go on the tension side. Wood would first show grief on the compression side, in the form of microscopic wrinkles on the surface, though it might hang in long enough that the tension side actually ruptures. For such materials, figures are often published for "flexural strength" and "flexural stiffness," which take into account the difference between tension and compression properties.

The way a bending stress is resolved into tension, compression, and shear depends on the geometric arrangement of the material, so strength and stiffness in bending is more a property of a structure than of a material. This is dealt with in chapter 2.

Specific strength

No matter what material a part is made from, you can obviously make it stronger by making it bigger. The problem with that approach is that the part becomes heavier at the same time, and for race cars and aircraft, weight is the enemy. What is needed is greater strength without a corresponding increase in weight. One of the things we're interested in, then, is the strength-to-weight ratio (what engineers call the *weight-specific strength*, or just *specific strength*) of a material. Comparing materials on this basis, it appears that modern composites are magic, and everything else is pure rubbish.

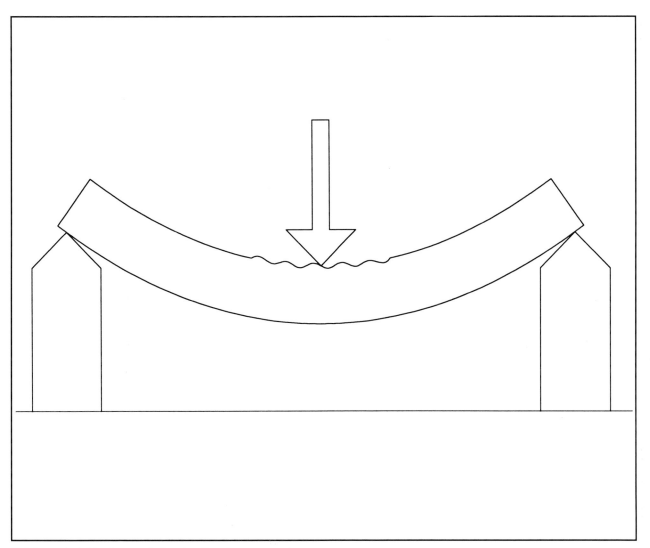

But if a material is stronger in tension than in compression,
the compression face will crush or buckle.

Strength and Density of Various Materials

	UTS ksi	Density lb/cu in	UTS/ Density
SAE 1010 Steel	60	0.3	200
SAE 4340 Steel	195	0.3	650
AZ31B Magnesium	42	0.065	646
SAE 2024 Aluminum	67	0.1	670
Spruce Wood	15	0.02	750
Nickel Superalloy	286	0.3	953
6Al-4V Titanium	170	0.165	1030
HM Carbon Fiber	325	0.07	4643
E-glass Fiber	525	0.09	5833
"S-2 glass" Fiber	692	0.09	7689
HS Carbon Fiber	600	0.07	8571
Aramid Fiber	435	0.05	8700

Yield Strength and Density

	Yield ksi	D lb/cu in	Y/D
SAE 1010 Steel	45	0.30	150
AZ31B Magnesium	32	0.06	500
SAE 4340 Steel	162	0.30	540
Spruce Wood	9.4	0.02	603
SAE 2024 Aluminum	66	0.10	653
Nickel Superalloy	265	0.30	880
6Al-4V Titanium	159	0.17	964
HM Carbon Fiber	325	0.07	4643
E-glass Fiber	525	0.09	5833
"S-2 glass" Fiber	692	0.09	7689
HS Carbon Fiber	600	0.07	8571
Aramid Fiber	435	0.05	8700

If the part is relatively short and squat, however, it may fail from pure compressive stress. Just how it fails will depend on whether its material is brittle or "ductile." If the material can yield, like common grades of steel or aluminum, it will squish like a stomped grape.

Composite materials are dealt with at length in chapters 8 and 9, but for now we should know that the really big numbers in the adjacent tables are for single strands of these amazing fibers. When the fibers are encapsulated in plastic resin, the resin adds weight but contributes almost nothing to strength. To compare them to other materials in usable form, then, these astonishing figures have to be reduced by about half. Impressive as such values remain, they only apply if all the fibers lie in the same path, and such a "unidirectional" composite has strength in only one direction. When several layers are built up with their strong axes at various angles, so as to give balanced properties in all directions, the layers which are not lined up with the imposed loads contribute little to the strength, but their weight remains.

Still, even when due allowance is made for these and other considerations (covered in the chapters on composites) that further reduce strength levels, composites remain enormously impressive from a specific strength point of view. Even plain old E-glass fibers—the stuff you have used, probably in cloth form, to patch a rusted fender—beats any metal by a factor of three to four. That is what makes composites so attractive for many race car (and aircraft) applications.

While the strength-to-weight properties of metals pale by comparison with composites, they have a number of practical advantages, including much higher service temperatures, their ability

to yield (as discussed above), and others (see Part II—Metals).

Among the metals, titanium, super alloy steels, and some non-ferrous super alloys are about tied for first place in specific strength. Their enormous strength, compared to more common metals, is not just a result of their exotic chemistry. Complex heat treatment techniques, and painstaking attention to surface finish and to maintaining the grain flow of the material are needed for the big numbers, so all these materials are challenging to machine, difficult to weld, and seriously expensive. And because their yield and ultimate strengths are so close together, they are extremely sensitive to notches, holes, and other abrupt changes in cross section that create stress concentrations.

Quite apart from their high cost, these considerations limit application of these materials to comparatively small, very highly stressed parts. Connecting rods, gears, and critical engine bolts are typical applications.

Of the remaining metals, the strongest aluminum alloys have 20–30 percent higher specific strength than heat-treated 4340 steel, and are four to five times better than everyday carbon steel. As metals go, magnesium fares poorly— even high-strength mag just barely edges carbon steel.

Perhaps surprising is the performance of plain old wood. Like unidirectional composites, however, wood is vastly stronger "with the grain" than across it. Also, wood changes size significantly with variations in humidity, is limited in its temperature tolerance, prone to decay, and is difficult to load directly in tension to anywhere near its real limits—the problem is grabbing hold of the ends in a uniform way. Nevertheless, it is astonishing to think that, on a weight specific basis, good structural wood surpasses heat-treated 4340 steel!

Stiffness

Strength and the ratio of strength-to-weight are not the only things that count in race car material selection. There is another important factor—stiffness. Strength, remember, is a measure of the amount of force required to break a sample of material. Stiffness, on the other hand, is an expression of how far the sample will deform under a given load—how "springy" it is. Thus, nylon rope is strong, but it extends noticeably under load—it lacks stiffness. Melba toast, on the other hand, is stiff, but weak.

While it is important that nothing in a race car actually breaks, it is often just as important that the deflections are kept small. Frames that lack stiffness cannot provide a stable platform for the suspension; gears that warp when heavily loaded lose their precise mesh, and so cost horsepower through extra friction; a springy crank throws off valve and ignition timing, and overloads the edges of engine bearings. Many race car and aircraft parts are "stiffness-critical"—if designed for tolerable deflections, they will more or less automatically have adequate strength.

The number representing stiffness is called the *modulus of elasticity*, or *Young's modulus*, and is represented by the capital letter "E" in materials handbooks. The figures typical for structural materials are in the range of millions or tens of millions of psi (a million psi is usually abbreviated to "msi"). But, if the strength is a few tens of thousands or hundreds of thousands of psi (a thousand psi is written as "ksi"), how do we get into the millions?

To illustrate, if our dangling 1sq-in bar started out, say, 100in long when unloaded, you would find (if it was made of steel) it would grow longer by about 0.003in for every thousand pounds you hang on the end. (Perhaps surprisingly, this same value holds true for all steels, from K-mart to NASA quality.) At that rate, a load of 30 mil-

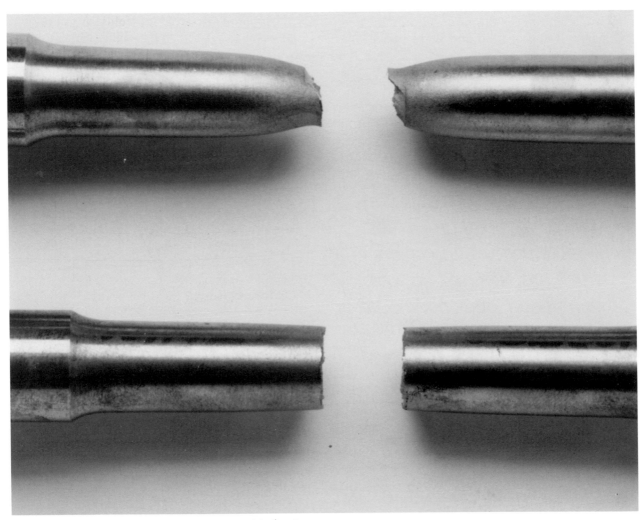

Ductile material, like mild steel, flows at loads well below the ultimate tensile strength. Brittle material—like steel heat-treated to highest strengths, or composites—fails at greater load, but without significant flow. *Gerard Gortzman*

lion pounds would apparently double the length of the bar. That value, 30 msi, is the Young's modulus of steel.

You could never apply that much load, of course—we have already seen that the yield strength of the very best grades of super alloy steel is no more than 300ksi, so the bar would extend (or *strain*, as engineers put it) just 1in before yielding. Nevertheless, the Young's modulus is the stress that would be required to double the original length of a specimen... *if* we could do it.

Comparing the stiffness of different materials reveals a very wide spread of values, and no particular pattern. The gap between composite materials and metals has disappeared. In fact, steel comes out looking pretty good compared to all but the high-stiffness carbon fibers. Metals that are "lighter" than steel (titanium, aluminum and magnesium) and aramid and glass fibers look sick, and wood is just a joke.

When several layers are built up with their strong axes at various angles, so as to give balanced properties in all directions, the layers which are not lined up with the imposed loads contribute little to the strength, but their weight remains.

Specific Stiffness

The situation is the same for stiffness as for strength—we can gain by using more material, but weight goes up at the same time. To compare the structural usefulness of different materials, we again need to take weight into account—thus the idea of the stiffness-to-weight ratio, or *specific stiffness*. As with specific strength, this is just the raw number for each material divided by its density.

When we compare materials on this basis, we run into perhaps the strangest coincidence in all of materials science—all common structural metals, E-glass fibers, and most kinds of wood have virtually identical values of specific stiffness in tension! This curious state of affairs has been described by Professor J.E. Gordon as "one of God's little jokes."

The implication of this is obviously that it is impossible to significantly increase the stiffness, for a given weight, of a simple strut or tie-bar loaded in tension by substituting among these materials.

Stiffness(E) and Density(D) of Various Materials

	E msi	D lb/cu in	E/D
Spruce Wood	1.4	0.02	70
6Al-4V Titanium	16	0.17	94
SAE 1010 Steel	30	0.3	100
SAE 4340 Steel	30	0.3	100
SAE 2024 Aluminum	10.6	0.1	106
AZ31B Magnesium	6.5	0.06	108
Nickel Superalloy	33.6	0.3	112
E-glass Fiber	10.5	0.09	117
"S-2 glass" Fiber	12.5	0.09	139
Aramid Fiber	16.3	0.05	326
HS Carbon Fiber	32	0.07	457
HM Carbon Fiber	100	0.07	1429

Chapter 2

Materials and Structures

The distinction between a material and a structure isn't always as clear as you might think. For example, steel is obviously a material, while a race car frame made from steel tubing is equally obviously a structure. But what about a single straight length of steel tube? Is it material or is it structure?

There is no correct answer to this, and if the load on the structure stresses the material in tension, it really doesn't matter—the strength and stiffness of the structure will depend simply on the properties of the material. This is a relatively rare state of affairs in the real world, however; it is much more common for material to be stressed in compression or shear as a result of the loads on the structure it comprises. In that case, how much the structure deforms and how much load it will take before failing will depend on its size

To get equal stiffness and strength with less weight, it is usual to rearrange the material so it is concentrated where it can do the most good.

and shape—on the way its material is arranged, in other words.

Imagine a beam, supported at both ends, and with a weight placed in the middle of its span. This will obviously tend to bend the beam which, as explained in chapter 1, will produce both tension and compression—the underside of the beam will be stretched, while its upper surface tends to shorten. If the beam has a constant depth over its entire length, these forces will be at a maximum right in the middle where the weight rests, tapering off to zero at the end supports.

The stress in the material depends on the size of the "bending moment"—the force produced by the weight, multiplied by the distance from the support—and so varies along the length. But the material stress also varies through the depth of the beam, changing smoothly from a maximum in tension right at the surface of the lower face, through zero at the middle of the beam (what engineers call the "neutral axis"), to a maximum in compression at the top surface. Thus, while the maximum bending moment depends only on the amount of weight and on the span, the actual peak stress in the material depends on the depth of the beam. Greater depth not only spreads the load over more material, it also gives the beam more "leverage" to resist the bending, so the material stress varies with the width of the beam and with the *square* of the depth.

Most of the material in regularly shaped beams is obviously underworked, so they are very

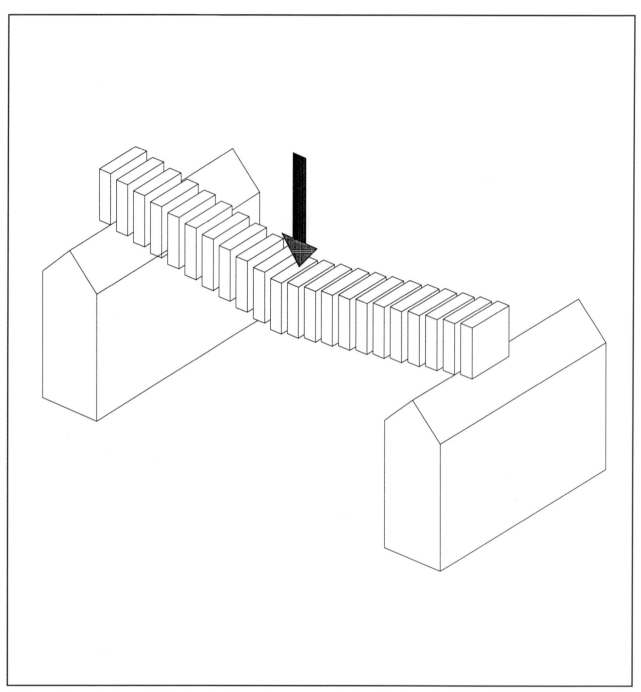

Quite apart from the forces that induce tension and compression into the surfaces of a beam that is subject to a bending load, there is a shearing component that acts uniformly, if the beam is regularly shaped, at every point along its length.

inefficient structures. To get equal stiffness and strength with less weight, it is usual to rearrange the material so it is concentrated where it can do the most good. The beam can be tapered, for instance, so it is thickest at the center, to equalize the bending moment throughout the length of the beam.

That, however, doesn't deal with the fact that the material at the neutral axis is just along for the ride. A common solution is to concentrate the material at the surfaces and remove as much as possible from the rest. That's the idea behind an "I" beam—two thick "flanges" separated by a thin "web," that does little more than hold the flanges in place. (Actually, the weight pressing downward will also produce a shearing force, equal at every point along the beam, that tries to slide each section past the adjacent section. The web also helps to resist this shear.)

What appears to be a "bending" load, we have discovered, actually breaks down into a combination of tension, compression, and shear. There are many other ways in which the forces acting on a structure become transformed by the shape of the structure into different kinds of forces in the material itself.

Imagine a simple frame made from tubes welded together into a rectangular box shape. If we firmly anchor one end, and start twisting at the other end, the rectangles formed by the faces of the box will all be distorted into diamond shapes. Obviously, if we replaced all the welded corners with ball joints, the structure would have no resistance at all to this twisting. It is only the welded joints that give our imaginary frame its stiffness. But if we weld a single tube running diagonally into each rectangular space, then to wrench the box out of square we would have to either stretch or shorten each tube. Apart from the fact that the box would suddenly have got vastly stiffer, something very interesting has happened—the torsional load on the structure as a whole has been converted into compression or tension loads in its parts. Whether the load in any given tube is compression or tension depends on which direction we twist the free end of the box.

Instead of a frame made from tubes arranged like the edges of an imaginary box, we could make a real box, say out of folded and riveted sheets of aluminum. Again, any effort to twist the box would tend to deform each rectangular face into a diamond shape. This time, though, we've converted the torsional load on the box not into tension or compression in its material, but rather into shear. These two examples should teach us that the loads in the material that a structure is made from aren't necessarily the same as the loads in the structure as a whole.

As we mentioned in chapter 1, parts with a stubby shape, like flat washers or the balls in a ballbearing, will fail in true compression—if the material is ductile, they will squash when the

yield strength of the material is exceeded; if it is brittle, they will fracture when the load goes beyond the ultimate strength. A slim part, though, will fail first by bowing or buckling.

Depending on which way the diagonal tubes run and which way we twist it, some of the tubes in our imaginary triangulated frame will be loaded in compression. Depending on the relationship between their wall thickness, diameter and length—what's called the "slenderness ratio"—they may fail in true compression (unlikely) or they may buckle. And the buckling could be local crinkling of the walls (this happens when you crush a beer can by stepping squarely on its end) or it could be an overall bowing of the tube. Depending, again, on wall thickness, diameter, and length, a tube overloaded in shear might also buckle (grip a beer can at each end and twist with all your might—see?), or it might fail in "pure" shear (try the same thing with an egg-roll).

The actual stress in the material as a whole can be quite low when this kind of buckling begins. The size of tubes needed for our imaginary frame is more likely to be set by their tendency to buckle or bow when the frame is twisted in one direction than by the likelihood of their failing in tension when it is wound the opposite way. Once a certain slenderness ratio is exceeded, in other words, it takes more material to carry a compression load without failure than to carry the same size load in tension.

For example, if we want to resist a 2,500lb tension load, using a steel rated at 50ksi, we would need a piece with a cross section of 0.05sq in—a rod just a shade more than 1/4in diameter would do fine. If we want to carry that load across a span of 2in, the whole tie-rod would weigh 0.03lb. Or if the ends of the rod needed to be a foot apart, then the rod would weigh exactly six times as much—0.18lb.

On the other hand, while we might get away with pushing on a 2in length of 1/4in rod with a 2,500lb force, there's no way we could do that if the rod were a foot long. It would, as we keep saying, fail by buckling, and to avoid that we would have to increase the diameter of the rod very substantially. So, while the amount of stuff—the weight of structure—needed to carry a tension load increases linearly with the distance over which the load is carried, structures resisting compression get heavier much faster than the span increases.

Interestingly, the weight of a solid pushrod increases as the square of the diameter, but the load to produce buckling increases as the fourth power of the diameter—double the diameter gives four times the weight, but sixteen times the buckling resistance, for the same material.

So if we have specified the diameter of a slender pushrod so it is just barely able to resist buckling under a certain load and then re-design for a load sixteen times greater, the weight of the rod only doubles. We have a structure that is eight times as efficient as before! The same thing

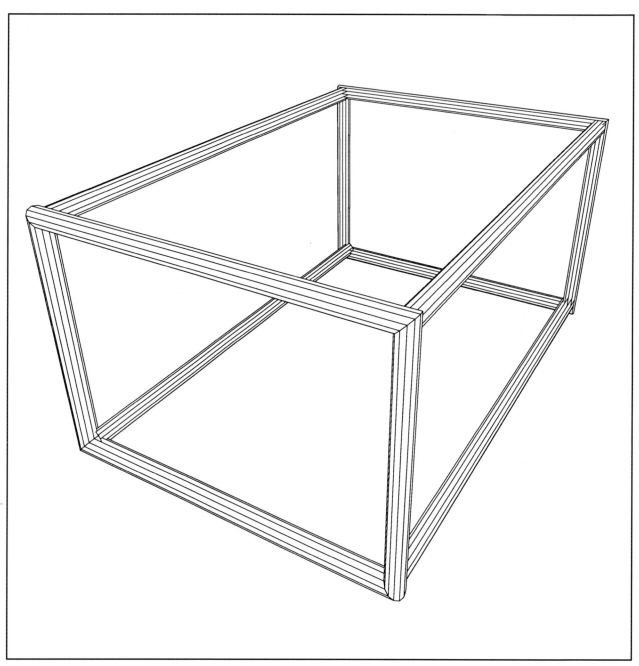

If a twisting force—a torque—is applied to a simple rectangular frame, the only resistance to that torque will come from the stiffness of the joints at the corners. Each flat face will distort into a diamond or "lozenge" shape.

happens in reverse as we reduce the size of the compression load—while the structure weight can be reduced, it can't be pared nearly as fast as the load diminishes, so the structure gets proportionally heavier as the loads get smaller. The worst case is obviously a very small load carried over a very large span.

Race cars are not made from long pushrods, but tubes, sheets, and thin plates respond by buckling in much the same way when pushed on

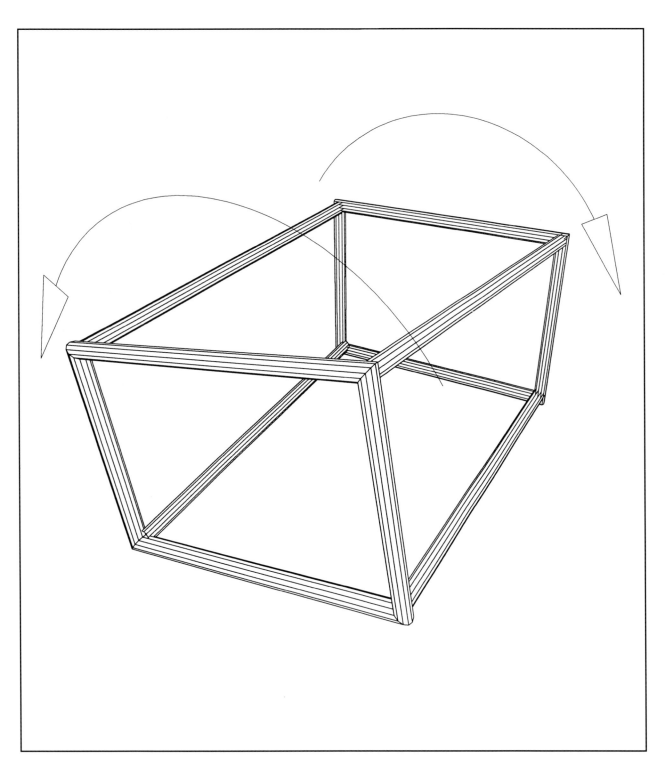

and also, as our aluminum box would show, when they are loaded in shear.

The resistance to buckling of a sheet or plate depends on the inherent stiffness of the material from which it is made, but it also depends on the thickness—in fact it is proportional to the *cube* of the thickness. A small increase in thickness therefore makes up for a sizable reduction in material stiffness. So, if two materials have different densities but the same ratios of strength and

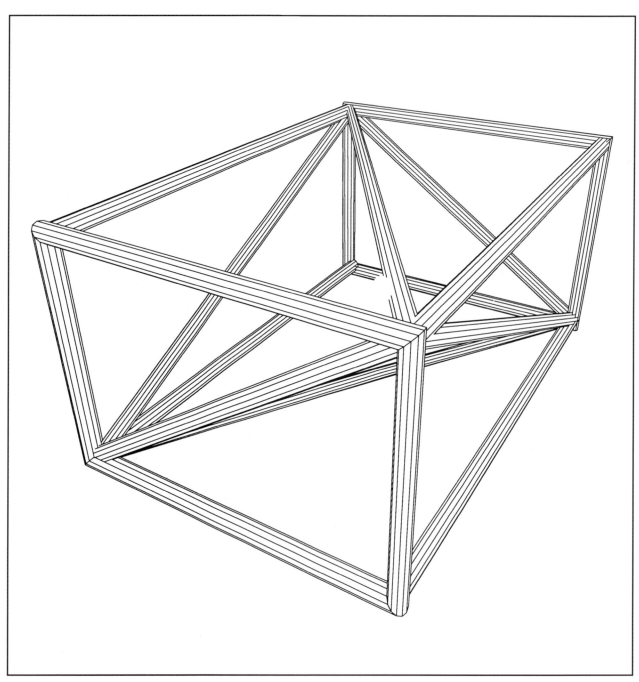

Breaking up each rectangular face into a pair of triangles prevents "lozenging" by converting the torque acting on the frame as a whole into tension or compression in each of the straight-line elements (say, tubes) that make it up.

stiffness to density, then any place where the potential failure takes the form of buckling, the less-dense material will make for a lighter part. Partly, that's why large, comparatively lightly loaded compression structures, like aircraft and race car monocoques, are made from relatively low density materials like aluminum, magnesium, composites, and even wood.

It may also help explain why, when dealing with structures where compression and shear

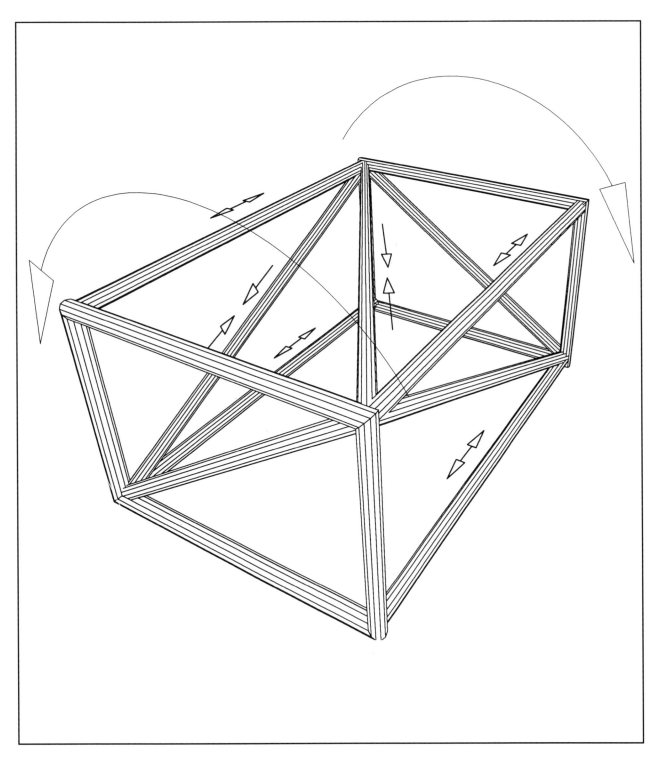

The stress in the material depends on the size of the "bending moment"—the force produced by the weight, multiplied by the distance from the support—and so varies along the length.

loads predominate and where stiffness and light weight are important, there is a general trend for the density of the material an object is made from to be related to the density of the object as a whole. Thus, most passenger cars are made mostly of steel; "high-density" aircraft (where the total weight is quite large compared to the total volume) are made of aluminum; "low-density" aircraft like gliders are often made from fiber glass; and rigid-framed airships (when they were around!) used a lot of magnesium. There is no end of exceptions to this pattern, of course, and over the range of weights and sizes of objects like automobiles, small boats, and light aircraft, comparable results have derived from virtually all

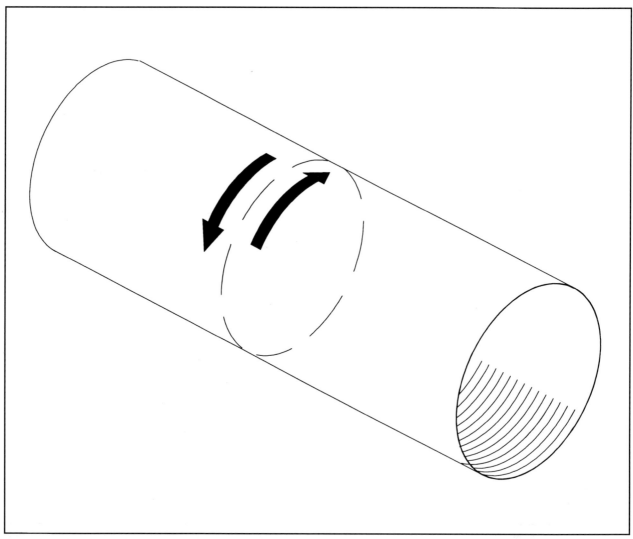

A torsional load applied to a tube (whether round or some other shape) turns into a shear force in the material.

"conventional" materials. Nevertheless, considering this relationship between the density of a component and the density of its material often suggests opportunities for weight saving.

Because different parts of a structure experience different kinds and sizes of load, it is tempting to use a variety of materials—each selected as being best for some particular task. This is done, of course, but apart from consideration of the widely different tools, facilities, and skills required to work with an assortment of materials, there remain the difficulties of joining, thermal expansion problems, and concerns about corrosion.

What appears to be a "bending" load, we have discovered, actually breaks down into a combination of tension, compression, and shear.

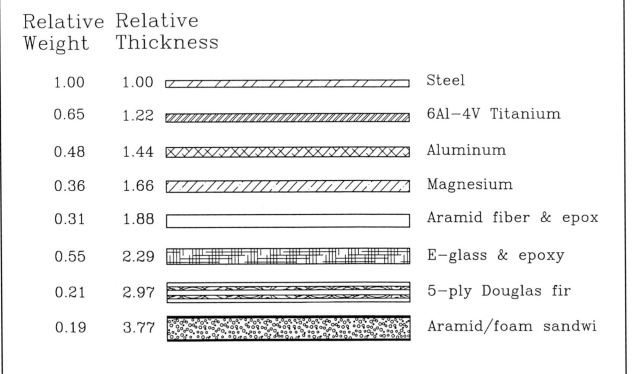

PANELS OF EQUAL BENDING STIFFNESS

Relative Weight	Relative Thickness		
1.00	1.00		Steel
0.65	1.22		6Al–4V Titanium
0.48	1.44		Aluminum
0.36	1.66		Magnesium
0.31	1.88		Aramid fiber & epox
0.55	2.29		E-glass & epoxy
0.21	2.97		5-ply Douglas fir
0.19	3.77		Aramid/foam sandwi

The bending stiffness of a flat panel depends directly on the stiffness of the material it is made from, and on the *cube* of the panel's thickness. For the same stiffness, a thick panel of low-density material can usually be made lighter than one made of a stiffer but denser material.

Chapter 3

The Real World

Fatigue

Fatigue in materials means much the same thing as it does in people: after working hard for a while, you get pooped out and can no longer develop full strength. Unlike people, though, materials don't recuperate after a rest.

When a structural material is loaded cyclically (that is, the load is applied, then removed, then applied again), its strength diminishes, so it will eventually fail under a much smaller load than it could have withstood when new. This problem is worst for parts that are very heavily loaded, and where the load turns on and off very frequently—connecting rod bolts, for instance, go through a stress-relax cycle every time the engine makes one revolution.

For some materials there is an "endurance limit"—a load that can be repeated virtually forever without failure. In the case of steel, that limit is about half the static strength. Aluminum is not like that; it gets weaker with every application of load. Still, a useful fraction of the original strength is retained for a very large number of load cycles. For most wrought aluminum alloys, the strength after 10,000,000 cycles remains around one-third of the static strength.

Of course, a load can cycle in many different ways. The stress in a part might vary from some maximum value in compression, through zero, to the same value in tension, then back again, and so on. Think connecting rod. This is called a "fully reversed" load, and is the worst case. Alternatively, the load might oscillate between zero and some maximum in tension, or between a small tension load and a big one. The proportion between the major and minor loads is called the "stress ratio," and the larger the stress ratio, the more severe the fatigue effects.

We can take advantage of this fact by pre-stressing the part, to reduce the difference between the maximum and minimum loads—the major load remains the same, but the minor load is increased, because the part never gets the opportunity to fully relax. That's the reason why

It is perhaps less generally appreciated that stiffening or "beefing-up" a part that flexes too much or has a history of breakage can worsen stress concentrations. . . .

Arranging for the smoothest possible transition at every change in cross section helps.

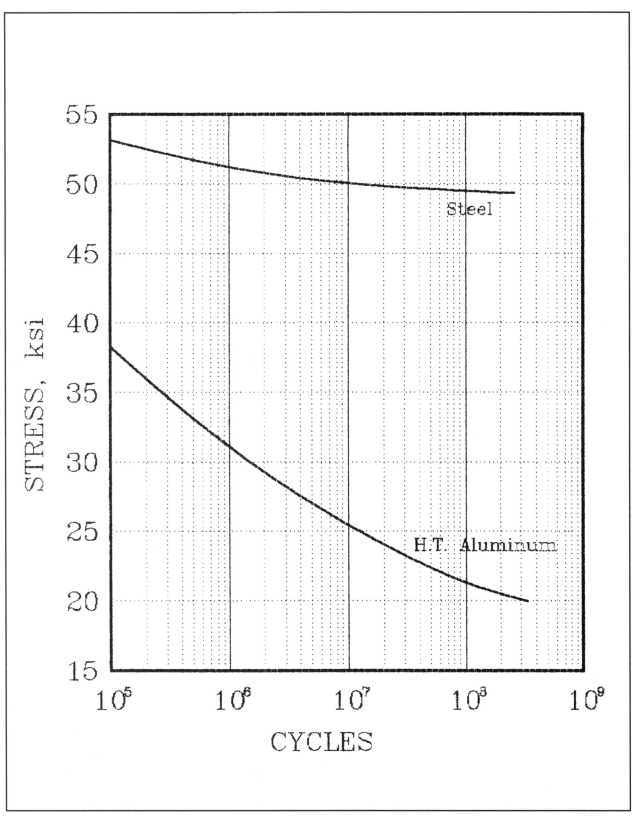

Initially, all materials become weaker with every application of load. In some, like steel, this "fatigue" seems eventually to level out—there is some cyclic load that they will carry indefinitely. In others, like aluminum, the strength continues to sink with every cycle.

many engine bolts are tightened within an inch of their lives on assembly, and why failing to torque them in this way leads to early failure.

While it seems paradoxical, somewhat the same effect can be achieved by having a part that is going to be cyclically loaded in tension start off with a pre-load in compression. If the load would produce a tensile stress alternating between, say, 0 (zero) and 60ksi in an untreated part, but the treated part begins life with a compressive pre-

load of 30ksi, in service the actual stress within the material would then vary between compression and tension. Normally, that would be a bad thing, but the peak stress in either direction has been cut in half, and the reduction in the maximum stress more than makes up for the fact that it now fully reverses.

Since fatigue failures almost inevitably begin at a surface, we don't need to arrange for a preload throughout all of a component—which would

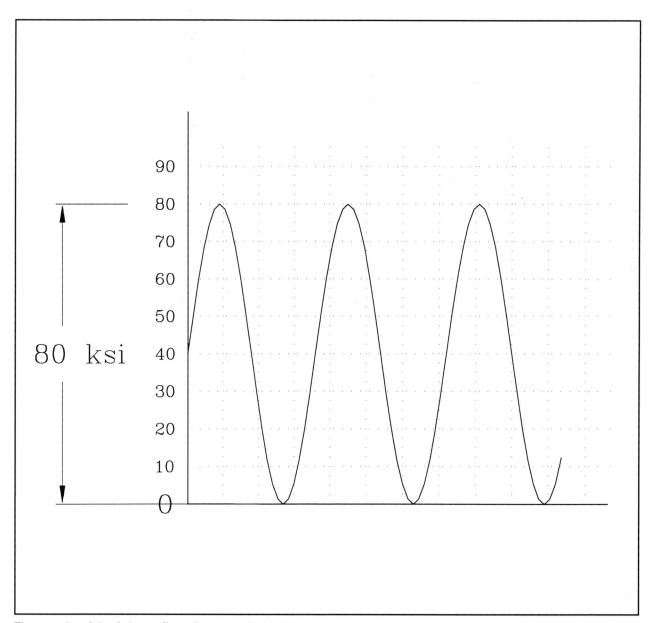

The severity of the fatigue effects from a cyclic load depends on the difference between the minimum and the maximum load experienced...

be impossible—but we can pre-load *part* of it. The fatigue life of aerospace bolts, for example, is greatly extended by introducing a residual compressive stress in the thread roots and the head-to-shank junction—the most critically loaded areas by a rolling operation. Shot peening is another way to achieve the same thing.

The fatigue characteristics of composites are generally superior to those of metals, but it is definitely *not* true that composites are completely immune to fatigue. Unidirectional composites based on carbon fiber lose perhaps one-third of their original strength after 10 million cycles of loading in tension; glass fiber composites are down to about half their static strength by that point. Unidirectional aramid laminates seem to do especially well in this respect, losing only about one-quarter of their static strength after 10 million cycles. They do much less well, however, under cycled bending loads.

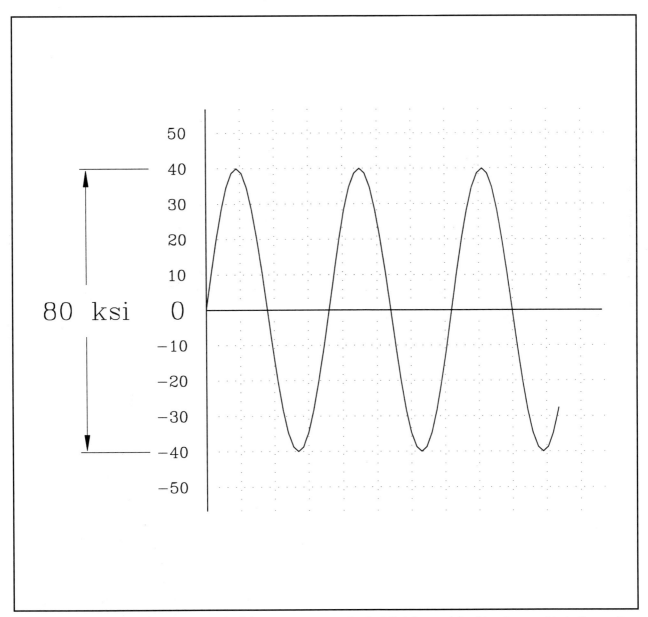

...so shifting the "zero" point by "pre-stressing" the part can return big dividends in fatigue life. Even though a "fully reversed" load (as shown here) is the worst kind for fatigue, the fact that the peak load has been cut in half more than makes up for that.

Stress Concentrations

In chapter 1 we described the concept of stress by talking about pulling on a 1in square metal bar with increasing amounts of force, until the bar broke. We said that if the bar withstood a force of 50,000lb but broke under a larger load, we could say the tensile strength of the material the bar was made from was 50,000 pounds per square inch (50ksi).

Now, if you drilled a 1/2in hole through the same bar, the cross sectional area right in line with the center of the hole would obviously be reduced to 1/2sq-in. You might reasonably expect that the bar could now hold a load of 25,000lb without fracturing. You would be wrong. The problem is not the arithmetic, it is the stress concentration produced by the hole.

If we gripped the original (undrilled) bar uniformly at its ends and pulled, the lines of force acting through it (what engineers call "stress tra-

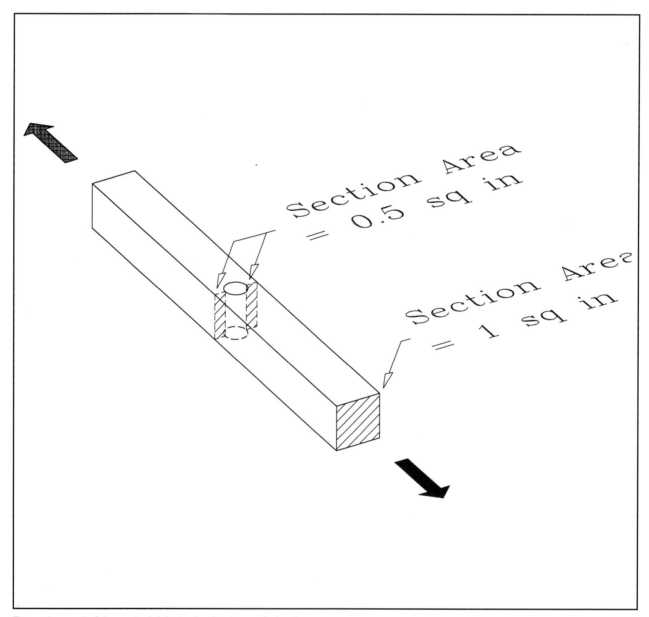

Removing part of the material that a load acts on obviously increases the stress. But apart from considering the "net section"...

32

jectories") would run in straight lines from end-to-end. But after we drill the hole, those lines of stress have to dodge around the hole, so they become crowded together—concentrated—so the local stress in the neighborhood of the hole is much higher than the average stress of 25ksi you might expect. In fact, the "stress concentration factor" for a round hole is three—the local stress right at the edge of the hole would be 75ksi. Bang.

A stress concentration doesn't need to be a hole. Any abrupt change in cross section—a notch, a groove, a crack, even a sudden *increase* in section area—focuses lines of stress into a small area. And the severity of the stress concentration depends on how suddenly the section area changes. (These things can be calculated, or you can look them up in tables found in engineering handbooks.) That's why crankshafts have smoothly rounded transitions between journals and webs; why high-strength aerospace bolts have large radius fillets at the head-to-shank junction; and why you don't find a lot of sharp corners in nature.

Although bolted connections are used in composite structures where unavoidable, and although some metal structures are adhesively bonded, metals are generally joined with bolts or rivets simply because joint strengths can be calculated with confidence, and because it is possible to get away with it.

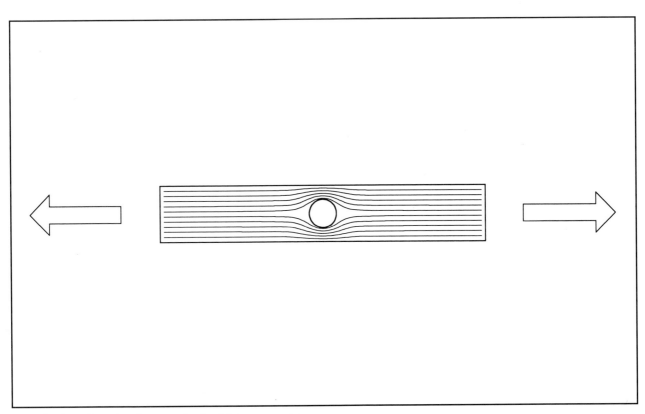

...we also have to make allowance for the "stress concentration" that results when the lines of stress are interrupted and crowded close together...

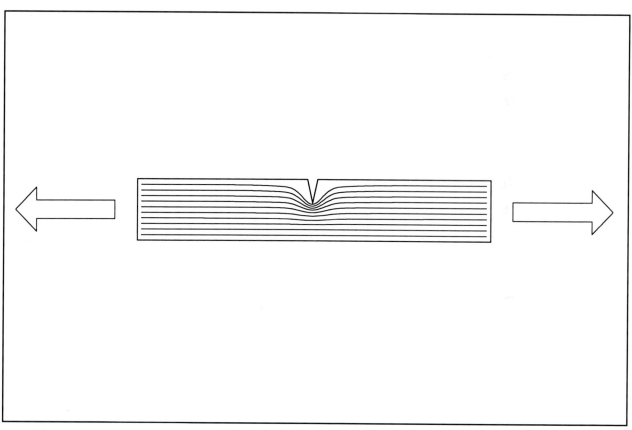

...and the sharper the change of section area, the worse the stress concentration becomes.

Most people who work around race cars understand something about stress concentra-

tions—that sharp edges and corners should be avoided, for instance. It is perhaps less generally appreciated that stiffening or "beefing-up" a part that flexes too much or has a history of breakage can worsen stress concentrations. Sometimes, for example, an improperly triangulated tube frame that has started to break up at the welded joints will be reinforced with a fillet plate welded across the troublesome corner. As often as not, the cure is worse than the disease, because the stresses are now even more localized at a small area on the tube wall where the fillet ends.

For all their evil consequences on static strength, stress concentrations have an even more severe effect on fatigue properties. Arranging for the smoothest possible transition at every change in cross section helps. For example, in a high-strength aerospace bolt, increasing the radius at the root of the thread by the most minuscule amount—from 14.4 percent to 18 percent of the thread pitch—increased fatigue life by nearly half.

Elongation is a calibration of the permanent increase in length of a tensile test bar at fracture—in effect, the two broken pieces are jammed back together and the total length then compared to the original length before testing.

> *At the atomic level, there is no real difference between mechanical stress and chemical stress. So, when a piece of material is taxed to very near its mechanical limit, it takes only the slightest bit of corrosion to push it over the edge.*

Corrosion

Lumps of iron, aluminum, magnesium, and other structural metals are not found in nature simply sitting around on the ground—their natural state is in combination with other elements. Refining separates the metal from the rest. Alas, the result is temporary, as the metal slowly reverts to the naturally occurring—and structurally useless—oxides, hydroxides and salts.

Most metals exposed to air, for example, react with oxygen, converting the surface into an oxide. In the case of aluminum, Titanium, and (so-called) stainless steel, this surface film is impermeable to air, so further oxidation is prevented. If the oxide film is damaged, it heals itself by the same process. (Anodizing is an electrical process that simply deposits a thicker, stronger coat of oxide on aluminum.)

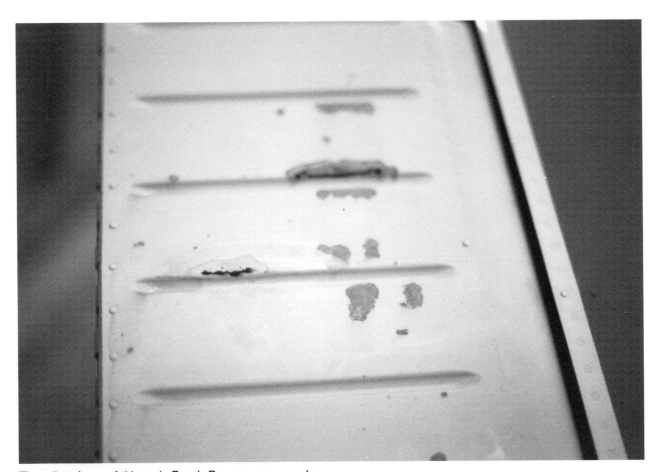

The tail surfaces of this early Beech Bonanza were made from magnesium. Despite protective zinc di-chromate primer and a coat of paint, corrosion has progressed rapidly in this most electrochemically active of common structural metals. *Forbes Aird*

The atmosphere, however, contains things other than oxygen—apart from other corrosive pollutants in the air, gasoline engines cough out sulfuric and nitric acid, dissolved in water vapor.

The oxide of iron (rust), on the other hand, forms as microscopic flakes that provide no protection to the underlying metal. Still, a coat of paint adequately defends iron against simple oxidation. Even chemically active magnesium oxidizes only slowly at room temperature. The atmosphere, however, contains things other than oxygen—apart from other corrosive pollutants in the air, gasoline engines cough out sulphuric and nitric acid, dissolved in water vapor.

Then there's *galvanic corrosion*. When dissimilar metals are connected and are wetted by an *electrolyte* (even contaminated damp air will do), a small electrical current flows through the electrolyte as the more active metal (the *anode*) gives

The alloying elements that give wrought aluminum alloys strength are arranged in micro-thin layers. Unfortunately, this "grain" provides a pathway for corrosion to work deep into the metal. Such "intergranular corrosion" is hard to spot at first, but the corrosion products occupy more volume than the parent metal, so eventually material flakes off. *Forbes Aird*

36

up electrons to the other (the *cathode*). At the same time, atoms from the electrolyte combine with the anode metal, creating a new material—crud! The cathode is unaffected. These electrical relationships are not limited to metals—carbon fiber is more cathodic than any common metal. So, while steel screws joining aluminum parts will corrode the aluminum, carbon fiber composites joined with steel or aluminum fasteners will remain unaffected, while the bolts rot away.

Electro-Chemical Potentials

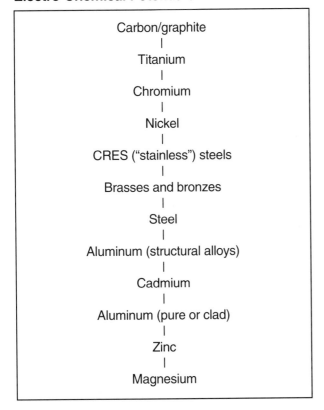

```
Carbon/graphite
       |
    Titanium
       |
    Chromium
       |
     Nickel
       |
CRES ("stainless") steels
       |
Brasses and bronzes
       |
      Steel
       |
Aluminum (structural alloys)
       |
    Cadmium
       |
Aluminum (pure or clad)
       |
      Zinc
       |
   Magnesium
```

It might seem that an easy fix for galvanic corrosion is to insulate dissimilar metals from each other. But all structural metals are alloys—2024 aluminum, for example, has discrete particles of copper distributed through it. Whenever the surface of an alloy is wetted by an electrolyte, these microscopic grains can serve as dissimilar electrodes to it—and not just on the surface.

Beginning at a pit at the surface, corrosion can penetrate an alloy by progressing along the grain boundaries. This *intergranular corrosion* is difficult to spot early on, but in its terminal stages it can manifest itself by *exfoliation*—thin leaves of metal flaking off. Some aluminum alloys and "stainless" steels are particularly susceptible.

While it is impossible to completely prevent a metal alloy structure from behaving as a battery, structural damage can be postponed for a very long time. Since it is always the anode that is consumed, we can provide a second, more active anode, which can be sacrificed to preserve the other, like the old trick of placing a copper penny on top of your car battery post. This is the principle behind *cladding*, in which a thin layer of chemically pure aluminum is bonded to a sheet of aluminum alloy. The protective oxide film on the pure aluminum surface combines with its self-sacrifice to make cladding convincingly effective. Similarly, when steel parts are plated with cadmium, the cadmium forms a stable oxide film of its own, but if the film is penetrated, the cad serves as a sacrificial anode for the steel beneath.

Not all corrosion is electrical. Bolted or riveted joints inevitably shuffle about, which gradually "frets" (wears away) some of the material at the contacting surfaces. *Fretting corrosion* eventually results in a wiggly structure, but when the joined parts are aluminum, there is an added twist. The aluminum oxide that is worn away is highly abrasive—it is used to make sandpaper. The result is a joint full of abrasive powder, revealed by black streaks around rivet and bolt heads.

At the atomic level, there is no real difference between mechanical stress and chemical stress. So, when a piece of material is taxed to very near its mechanical limit, it takes only the slightest bit of corrosion to push it over the edge. The outcome of this combination of chemical and mechanical stress is called *stress-corrosion cracking*, or *stress-fatigue cracking* when the mechanical loads are fluctuating.

The proportion between the major and minor loads is called the "stress ratio," and the larger the stress ratio, the more severe the fatigue effects.

Toughness & Damage Tolerance

We have already spoken of the ability of metals to yield or permanently deform without breaking. This virtue is discussed further later, but it is worth repeating that this helps make metals "safe" materials—they give signs of distress before complete failure and, to some extent, they can redistribute local overloads resulting from stress concentrations. This latter characteristic is part of what is meant by toughness. (Unfortunately, the greater the strength we achieve from any given metal by heat treatment or cold-working, the less yielding it generally exhibits, so the less tough it is.)

The ability to yield is not strictly necessary to provide toughness. Composites and wood—nature's own composite—are also tough materials in the above sense, even though they cannot yield like metals. The way such fibrous materials deal with stress concentrations is described in chapter 8.

One gauge of toughness is the impact energy required to fracture a sample, with or without a notch cut in its surface—as measured by Charpy and Izod tests. In this respect, glass and aramid composites are roughly tied—both are perhaps five times better than carbon fiber composites, aluminum, or magnesium. Most structural and machinery steels lie somewhere in between, depending on alloying and heat treatment. For comparable strength levels, titanium does somewhat better than steel in these sorts of tests. Beryllium, conversely, has a nasty reputation for crack sensitivity.

In another sense, toughness reflects the total amount of energy soaked up when a structural el-

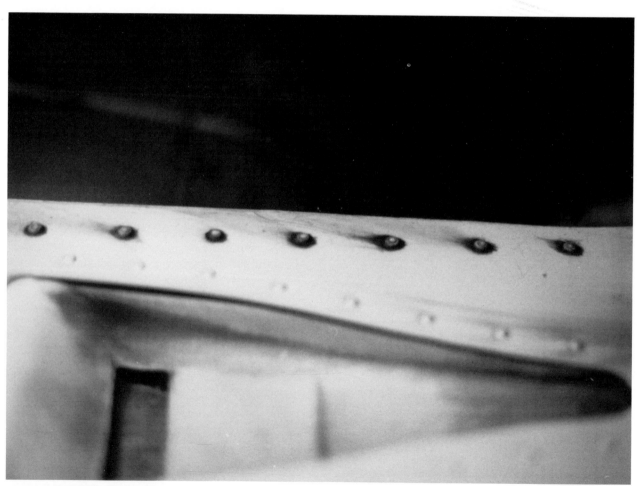

"Fretting corrosion": The shuffling of bolted or riveted joints wears away a little metal. When the metal is aluminum, the powder immediately forms aluminum oxide. Aluminum oxide is the black/gray stuff that some sand-paper is made from—it is highly abrasive. *Forbes Aird*

ement fails. By this measure metals and aramid composites (which exhibit ductility similar to metals when overloaded in compression) show up better than the stronger but more brittle carbon fibers. Glass fibers are better than carbon fibers in this regard, though not as good as aramid or metals.

The ability to yield is not strictly necessary to provide toughness.

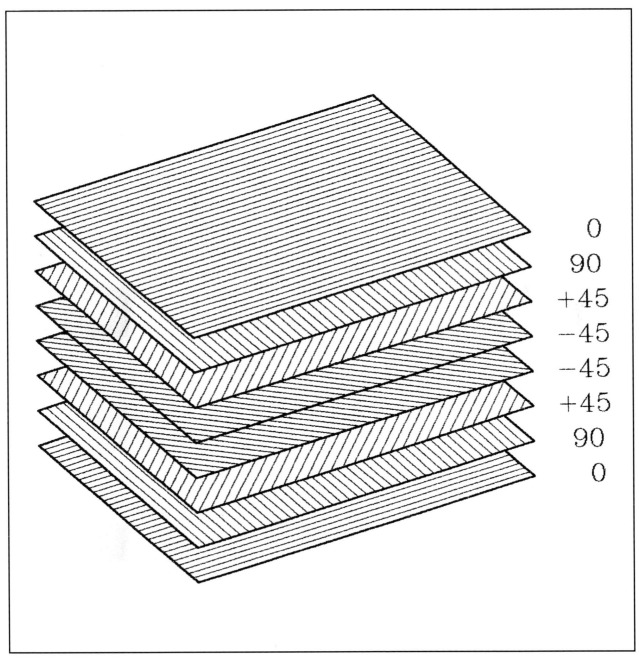

0
90
+45
−45
−45
+45
90
0

To prevent bolts or rivets from splitting a composite made up from multiple layers of unidirectional material, extra material has to be added at 45 and 90 degrees to the principal load, which adds weight fast. That's one reason why composite structures are usually adhesive bonded.

Since fatigue failures almost inevitably begin at a surface, we don't need to arrange for a pre-load throughout all of a component—which would be impossible—but we can pre-load part of it.

While they are not a direct measure of toughness in either of the above senses, figures for "elongation" and/or "reduction of area" are often quoted for metals, and for any given metal alloy there is usually a correlation (though not a linear one) between toughness and the elongation or reduction of area specs shown by an alloy in various states of heat treatment. Elongation is a calibration of the permanent increase in length of a tensile test bar at fracture—in effect, the two broken pieces are jammed back together and the total length then compared to the original length before testing. Reduction of area expresses the amount of necking-down of the test bar. Both are expressed as a percentage change.

Joints

The ability of some materials to redistribute excessive local stresses comes in handy around bolted connections and fittings. While such point loads always make a structure heavier than the theoretical minimum, more additional material has to be provided around bolted joints in carbon fiber than in glass; aramid and metals need least of all.

For instance, a carbon fiber laminate optimized for mechanical fasteners (bolts or rivets), would have to have 60 percent of the fibers around a bolt hole running at plus-and-minus 45 degrees to the load direction; aramid fiber laminates require only half as much. Although bolted connections are used in composite structures where unavoidable, and although some metal structures are adhesively bonded, metals are generally joined with bolts or rivets simply because joint strengths can be calculated with confidence, and because it is possible to get away with it. There is something reassuring about big fat bolts.

The ability to use rivets and bolts also has ramifications for inspection, maintenance and repair. Any fool can see a sheared rivet or a missing bolt, but who can see a defective adhesive bond? And assuming you could see it, how would you fix it? There is much to be said for "erector-set" construction when it comes to dealing with localized defects or damage.

Introduction

As discussed in chapter 1, all materials change shape when they are loaded, and all materials break when the load becomes too large. Unlike almost any other material, however, metals have the ability to respond to overloads by changing their character from an "elastic" solid that flexes under load, then springs back completely when the load is removed, to a "plastic" substance that flows like putty. (The only other material that exhibits this property is aramid fiber, and then only when it is loaded in compression—see chapter 9.)

This property not only makes it possible to form flat sheets of steel or aluminum into complex shapes like fenders, it also allows metals to accommodate local overstressing by "running away" from the load. Imagine a pattern of four bolts, for instance, where one is a slightly better fit than the others. That one bolt will initially carry all the load, until the material springs enough for the other bolts to take up a share.

If the material is perfectly elastic, however, the share will never be equal, so beyond a certain point something will snap, and you'll be left with three bolts. On the other hand, if that one bolt or the material surrounding its hole can flow, the overloaded area will permanently change shape, re-distributing the stresses, and forcing the material at the other three attachments to take up a fairer share of the load.

In both these respects—formability and relieving local overloads—yielding is a virtue. But there is a flip side. Yielding means that the part has permanently changed shape, so if yielding progresses much beyond this "load-equalizing" stage after a piece is formed to the desired shape, in most cases we have to regard the part as failed, as suggested in chapter 1. That means that, when dealing with materials that have an identifiable yield point, we can seldom use ultimate strength as a basis for design.

Ironically, metals that are modified by cold-working or heat-treatment to give higher strength invariably exhibit less ductility—less ability to yield. And composites and the other in-

---

The rolling or drawing operations which transform a cast ingot into sheets or "long stock" (tubes, angles, etc.) improve strength by work-hardening and by helping to form an organized grain structure within the material.

---

credibly strong man-made materials completely lack this ability. This may help explain the need for extreme attention to detail, including precision fits, when dealing with very high-strength materials, and should make clear that a stronger material doesn't always make for a stronger structure!

Metals can be bought either as castings or as "wrought" (shaped) products—bars, plates, sheets, tubes, angles, and such. Often a shape is

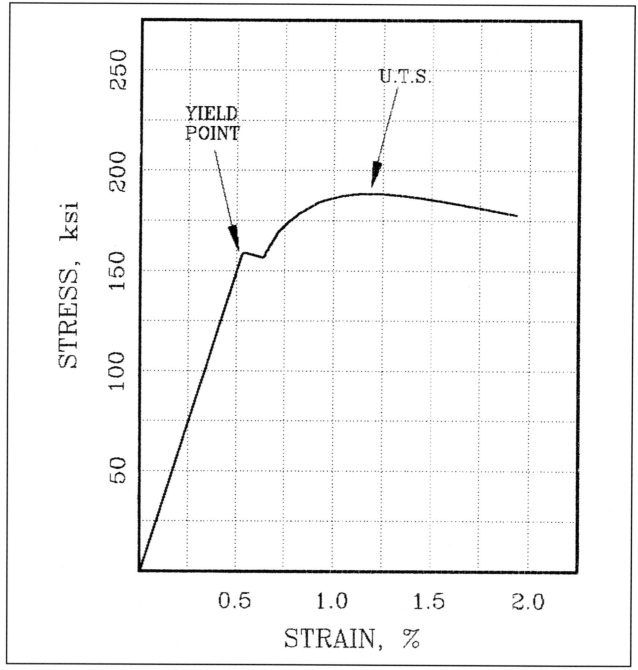

Like most structural materials, metals will "spring" elastically when loaded, then spring back when the load is removed. Uniquely, metals do not immediately rupture if the load gets too large—they turn to putty and yield. Although the greatest strength is achieved *after* the yield point, the metal has permanently changed shape, so must be considered "failed."

needed which cannot be wrought (how would you roll or otherwise form an engine block?), so you're stuck with a casting. But many race car parts are fabricated from wrought shapes—tubing for instance. The rolling or drawing operations which transform a cast ingot into sheets or "long stock" (tubes, angles, etc.) improve strength by work-hardening and by helping to form an organized grain structure within the material. Note that the degree of improvement depends partly on whether this shaping is done hot or cold.

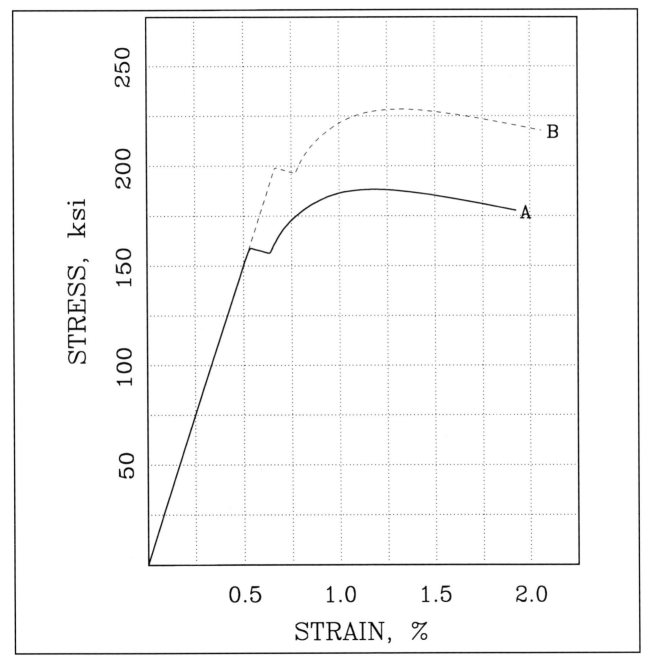

Heat-treating or cold-working metals (dotted line B) can greatly increase their strength, but the yield strength increases faster than the ultimate strength, so the material has become more brittle and thus in some senses less safe.

Chapter 4

Iron and Steel

About three-quarters of the weight of a production passenger car is accounted for by "ferrous metals"—iron or steel. In purpose-built race cars, the proportion is much lower, a trend that began with the advent of high-strength aluminum alloys in the thirties and that has recently accelerated, following the widespread availability of modern composites. Yet if we had only one structural material to work with, it had better be steel, or making automobiles would be impossible. While you can make a mighty efficient frame out of aluminum, and while composites do a better job than metal in many applications, when stresses are both complex and very tightly focused, or the temperature is high, steel is pretty hard to beat. Indeed, it is difficult to contemplate making a crankshaft or a wheel-bearing out of any other material.

With a density of 0.29–0.3lb per cubic inch, steel is the heaviest conventional structural material. Still, for parts that are sized by the loads they have to carry (such as bolts, gears, and spindles), high-strength steels show up remarkably well, because of their strength-to-weight ratio. Of course, you only get 300ksi steel from heat-treated aerospace grades—the sort of material premium rod bolts are made from. Lawn furniture is made from lesser stuff.

Ultra-high-strength alloys are discussed in chapter 11. Here we will consider stuff that is familiar, weldable, and can be machined, yet which is tough and fatigue-resistant, and has about as much stiffness in relation to its weight as any

Numbering System for Steel Alloys

Alloy number	Major alloying element(s)
13xx	Manganese
23xx, 25xx	Nickel
3xxx	Nickel-Chromium
40xx	Molybdenum
41xx	Chromium-Molybdenum
43xx, 86xx, 87xx, 93xx	Nickel-Chromium-Molybdenum
46xx, 48xx	Nickel-Molybdenum
5xxx, 5xxxx	Chromium
61xx	Chromium-Vanadium
92xx	Silicon-Manganese

other structural metal—carbon steel and conventional alloy steel. (Carbon and iron together make up an "alloy" called steel, but most people call this metal "carbon steel" or just plain "steel." The term "alloy steel" is reserved for cases where the *principal* alloying elements are other than carbon.)

Chemically, there's not a lot of difference between one grade of steel and another, nor even between steel and iron, for that matter. What you have got is almost all iron (95 percent and up, in most cases), with just a dash of carbon and usually a touch of manganese and a few other elements. Carbon content is the basic difference between iron and steel: while pure iron is a soft, silvery lab curiosity of virtually no structural value, a fraction of a percent of carbon makes a world of difference. Casting iron has more than 2 percent carbon; steel contains less than that.

The Workhorse—Carbon Steel

If a large amount (somewhere around 0.9 percent) of carbon is added to iron, and the resulting alloy is "quenched" from a high temperature by plunging it into cold water, strength levels over half a million psi can be obtained. There are a couple of problems, though. First, the effect can only be achieved in very thin sections, like piano wire or razor blades, because the interior of thicker sections cannot be chilled fast enough to achieve the rapid solidification necessary for such extreme hardness. Such thin sections obviously have a very large surface area in relation to their volume, and so are very much affected by corrosion, but there is another, larger problem. With such "file hard" high-carbon steel, the yield strength is indistinguishable from its ultimate strength—it is impossibly brittle, in other words.

On the other hand, when you buy a "no-name" piece of steel plate, bar, or tube from a small metal dealer, what you likely get is ordinary low-to-medium carbon "mild" steel, containing somewhere in the range of 0.1–0.3 percent carbon, with a yield strength of somewhere between 30 and 70ksi, depending partly on how much carbon is present and partly on whether the forming of the metal into plate, bar, or tube was done hot or cold. Plain carbon steels are identified by a four-digit number starting with "10." The next two digits indicate the percentage carbon—1015 steel, for instance, contains 0.15 percent carbon. A second digit other than "0" means the steel contains other elements. "1110," for instance has added sulfur; "1115" contains sulfur plus phosphorous. These elements are added to make the steel "free-machining" for the economical manufacture of low-strength screw machine parts, but produce comparatively weak and brittle parts. While steel made stronger and harder by cold working or by upping the carbon content provides a smaller safety margin, anything with less than about 0.25 percent carbon just can't provide sufficient strength for racing applications, and within the range of 0.05–0.3 percent carbon, there's usually plenty of ductility anyway.

This is one of the great beauties of mild steel—if you overload it, it will stretch out of shape, but it takes quite a bit more to actually break it, and in the meantime it has been giving some very clear indications that it is hurting.

When the yield strength gets above, say, 60 percent of the ultimate strength (as the brittleness increases, in other words), misfits and stress-concentrations like nicks, notches, and sharp corners become much more critical. A stronger material doesn't always make a stronger structure!

Alloy Steels

One way to have your cake and eat it too is through the use of alloy steels. There, you can get yield strengths much higher than with plain carbon steel, and still maintain a decent spread between yield and ultimate strengths, returning some of the early-warning and damage-tolerance benefits of a ductile material. Raising the strength level of alloy steels by the use of heat-treatment causes them, too, to become more brittle and less forgiving, but the whole trade-off is moved up the scale several notches compared to carbon steel, and the heat-treating process provides the opportunity to tailor this trade-off point very closely. Carbon content affects the balance between ductility and strength of alloy steels in much the same way as plain carbon steel; bearing balls, for example, contain just over 1 percent carbon (in addition to chromium), and achieve enormous strength (well over 300ksi), but ab-

Raising the strength level of alloy steels by the use of heat-treatment causes them, too, to become more brittle and less forgiving, but the whole trade-off is moved up the scale several notches compared to carbon steel, and the heat-treating process provides the opportunity to tailor this trade-off point very closely.

solutely no ductility.

Like carbon steels, alloy steels are also specified by a four-digit number. Again, the last two digits refer to the carbon content; the first two usually indicate the type and percentage of the principal alloying ingredient(s). Some of the more popular alloying elements are manganese, nickel, chromium, and molybdenum (these last two together make "chrome-moly" or "chromoloy").

Chrome-moly steel was popularized years ago by the aircraft industry, which wanted an alloy for sheet and tube production that would be reasonably strong yet formable and weldable in normalized (soft) condition, and which could be readily heat treated to higher strengths, while still retaining decent ductility. One chrome-moly alloy much used by racers is 4130 (the "41" stands for a chromium-molybdenum alloy, while the "30" designates 0.3 percent carbon), which can return yield strengths of over 130ksi with reasonable ductility, and more if some loss in ductility can be accepted. Since welding is a mighty handy way to join parts, but trashes a heat-treat, it is comforting to know that 4130 delivers 90ksi in normalized condition.

4130 has its limits, though, and where higher strength and/or greater toughness is needed, especially in thicker sections, an alloy containing nickel is often chosen; 4340 is one such. In the normalized condition, 4340 returns a yield strength of maybe 75ksi and a UTS of 100ksi. Properly heat-treated, 160ksi yield and a UTS greater than 180ksi is available, even in large-diameter bars.

Another high-alloy steel is 5140. It shows somewhat inferior strength compared to 4340, with both ultimate and yield strengths down by 15–20 percent. 5140 has no nickel or molybdenum in it, but a healthy dose of chromium makes it possible to "nitride" the journals for a super-hard wearing surface, which may account for its popularity with makers of aftermarket crankshafts.

Steels containing 3.0–3.75 percent nickel are widely used for medium-to-high strength bolts and other hardware. 2330 steel, with 3.5 percent nickel and 0.3 percent carbon, for example, is the material much light aircraft hardware is made from. For such applications it is heat treated to a condition that still provides ample ductility with a UTS of 125ksi.

Conclusion

Steel is very dense, but high-strength alloys are readily available which make it strong in relation to its weight. It is relatively cheap, it resists fatigue well, and most racers have the facilities and skills to weld it or machine it. Where stiffness is more important than strength, then for metal parts in pure tension you may as well use a cheap grade of mild steel, since no metal—not alloy steel, not aluminum, not even titanium—can significantly improve on its specific stiffness. Where the highest strength is needed for a given size (as opposed to weight), high-strength alloy steel can be challenged only by titanium and high-strength composites.

Some Alloy Steels

Alloy	Heat Treatment	UTS ksi	Yield ksi	Elongation %
SAE 1018	Cold drawn	82	70	20
SAE 1035	Tempered @ 900 F	93	63	
SAE 2330	Water quenched @ 1550 F Tempered @ 1000 F	121	95	–
SAE 2340	Oil quenched @ 1550 F Tempered @ 900 F	147	127	–
SAE 4130	Hot rolled, annealed	86	56	29
SAE 4130	Water quenched @ 1550 F Tempered @ 1000 F	146	133	17
SAE 4340	Oil quenched @ 1550 F Tempered @ 900 F	195	175	15
SAE 5140	Oil quenched @ 1550 F Tempered @ 900 F	160	140	–

'Stainless' Steels

The tendency of iron and steel to rust can often be countered by protecting the exposed surfaces with a layer of plating or paint, but there are some applications where an inherently more corrosion-resistant material makes better sense. When more than about 11–12 percent chromium is added to steel, the resulting material shows excellent resistance to many forms of corrosion, especially at elevated temperatures. The protection is provided by the chromium, which combines with oxygen from the atmosphere to form a microscopically thin but impervious layer of chromium oxide on the surface. This "stainless steel" is thus widely used for everything from household cutlery and kitchenware, to marine fittings, to aircraft structures and fasteners.

In fact, "stainless" actually isn't stainless at all; if the other conditions for galvanic corrosion are met (see chapter 3, "Corrosion"), it will rust away merrily in some severely corrosive environments, such as salt water. The aircraft industry more correctly refers to this stuff as "corrosion resisting" or "CRES" steel.

Whether you call it "stainless" or "CRES," the most common grade is the "300" series, containing chromium and nickel. 30302 stainless (302 for short), widely used for CRES airframe bolts, is typical. With a nominal composition of 18 percent chromium and 9 percent nickel, together with 0.15 percent carbon, it is good for about 35ksi yield, 85ksi UTS in the annealed condition, and has excellent ductility and formability. It cannot be heat-treated, but forming operations like tube drawing cause work hardening, which can raise the ultimate strength to as much as 200ksi, though there is negligible ductility at this strength level. 125ksi UTS, with yield strengths to 95ksi, are more typical values for a moderately work-hardened 302 alloy.

The 300 series CRES steels are readily weldable, though the effect of heating in the range of 900–1500 degrees F causes the metal to become "sensitized," with an adverse effect on resistance to corrosion—especially stress-corrosion cracking. To avoid this problem, some grades (like 30321) are available which contain columbium and titanium to "stabilize" against this sensitization. 300 series stainless is non-magnetic.

Using a magnet to identify "stainless" has its limits—there are various straight chromium CRES steels (with no nickel) which *are* magnetic. They form the "400" series, some of which (the "martensitic" grades) can be heat-treated up to 200ksi UTS, but which sacrifice corrosion resistance. 40416, with 13 percent chromium and not more than 0.15 percent carbon, is representative of these. Other nickel-free CRES alloys in the 400 series are "austenitic"—like the 300 series they cannot be heat-treated, but do work harden. Their long suit is improved corrosion resistance at high temperatures.

Finally, there are precipitation hardening (PH) stainless steels. 17-4 PH, for instance, with 16.5 percent chromium, 4 percent nickel, and 4 percent copper, can provide strengths comparable to more familiar heat-treated alloy steels (say 180ksi UTS), yet unlike them it is castable. This is the stuff used to make "stainless" aftermarket rocker arms for race engines.

Cast Iron

"Casting" iron is generally defined as iron containing more than 2 percent carbon, though practical grades of castable ferrous metals also contain significant amounts of silicon. "Detroit Wonder Metal," as it is sometimes called, has been widely disparaged as weak, brittle, and uselessly heavy stuff, and in its cheapest and nastiest form, it is. While it is always heavy, cast iron can also exhibit surprising combinations of strength and ductility, together with other desirable properties, and we have already seen that high density alone is not a reason to condemn a material. The possibility of producing large parts of complex shape by casting, rather than machining from a solid lump, is just one of a number of virtues possessed by cast iron.

Although the carbon content of cast iron can vary from under 2 percent (in which case it is, technically, not cast iron at all!) to more than 4 percent, the different properties displayed by various forms of cast iron have more to do with how the carbon is dispersed through the iron than with the actual amount present. If all the carbon is combined with iron, in the form of iron-carbide particles, the result is incredibly hard and brittle stuff called "white iron"—a fine example of an untrustworthy material, despite its high ultimate strength, which can exceed 160ksi. Alternatively, there may be some free carbon in the form of tiny flakes. That results in "gray iron." ("White" and "gray" are so called because of the color of a freshly fractured surface.)

Gray irons are often designated on the basis of their tensile strength, thus No. 35 provides

35ksi, No. 50 gives 50ksi, and so on. 55ksi UTS or more is achievable but, as you might expect, the ductility of gray iron generally diminishes with increasing tensile strength, though certain controls over the casting process (originally developed by Meehanite Metal Corp.) give wide flexibility in tailoring the finished material properties. Gray iron is widely used for engine castings and brake rotors.

By gradually heating white cast iron to very high temperatures (as much as 1700 degrees F) over a period of a couple of days, then allowing it to cool to room temperature even more slowly, some of the carbon falls out into irregularly shaped lumps, yielding "malleable" iron with yield strengths up to 35ksi combined with an exceptional ductility—elongation can be as much as 18 percent. By carefully varying the rate of cooling of various parts of an iron casting through the use of "chills" in the mold, a part can be produced with white iron at the surface, for exceptional hardness and wear resistance, while the more slowly cooled core takes the form of gray iron. This technique is used to produce many camshafts and lifters.

Better combinations of strength and ductility throughout a casting can be attained by the addition of small amounts of magnesium or cerium, which together with special process controls, causes the free carbon to take the form of tiny spheres—hence "spheroidal graphitic cast iron," otherwise known as "ductile iron." Ductile iron is specified by three numbers, separated by dashes, which reflect the UTS, yield, and elongation, respectively. Thus, 60-45-10 has a UTS of 60ksi, a yield strength of 45ksi and 10 percent elongation. If 3 percent elongation is acceptable, there is an 80-60-3 grade.

As with steel, the addition of other alloying elements can significantly improve both strength and ductility, and small amounts of nickel and/or chromium and/or molybdenum are often added to all types of cast iron. Also as with alloy steels, the proportion of these alloying metals is small—typically less than 3 percent, but "high alloy" irons may contain ten times that much.

For the exhaust turbine housing of turbochargers, for example, a high alloy iron ("Ni-Resist D2") is widely used. Comprising 20 percent nickel, 2.5 percent chromium, 2.25 percent silicon, and 0.95 percent manganese, together with not more than 3 percent carbon, "Ni-Resist D2" has a UTS of 58–60ksi, retains useful strength to very high temperatures, and is strongly corrosion resistant. Very large amounts of chromium are also sometimes added to the iron used to make motorcycle disc brake rotors for no better reason, as far as we can tell, than to prevent an ugly rusted appearance after a night parked in the rain.

Chapter 5

Light Metals

Aluminum

As we suggested in chapter 1, strength and stiffness for a given weight are important criteria for high-performance materials. We saw there, however, that there is not much to choose among common structural metals on the basis of specific strength, and virtually no difference in terms of specific stiffness.

On the other hand, we argued in chapter 2 that when there are compression loads that tend to make a part buckle, and when the various candidate materials have the same specific stiffness and strength, a less dense material will always save weight. That fact quickly led to the adoption of aluminum for airplane construction, and to the development of a wide assortment of structural alloys.

Just as with steel, the addition of small amounts of other metals to aluminum results in alloys with much greater strength than the pure metal and, also like steel, many of these alloys can be heat treated or "work hardened" (by rolling or other forming operations) to even higher strength levels. After decades of confusion and overlapping specifications, wrought aluminum alloys are now identified by a four-digit number recognized throughout the English speaking world. In each case, the first digit indicates the major alloying ingredient. Heat treatment is denoted by the letter "T," followed by one or more digits. In the case of non-heat treatable alloys, the improved strength properties resulting from work hardening are identified by the letter "H,"

Numbering System for Aluminum Alloys

Alloy number	Major alloying element(s)
1xxx	99% Pure Aluminum
2xxx	Copper
3xxx	Manganese
4xxx	Silicon
5xxx	Magnesium
6xxx	Magnesium & Silicon
7xxx	Zinc
8xxx	Other elements

followed by two or more digits. "2024-T4," for example, indicates a heat treated alloy from the copper alloyed "2000" series.

After decades of confusion and overlapping specifications, wrought aluminum alloys are now identified by a four-digit number recognized throughout the English speaking world.

Some Wrought Aluminum Alloys

Alloy	HT	UTS ksi	Yield ksi	Elongation %
SAE 1100	as welded	13	5	35
SAE 1100	H12	16	15	12
SAE 2024	T4, T351	68	47	20
SAE 2024	T361	72	57	13
SAE 3003	O	16	6	30
SAE 3003	H18	29	27	4
SAE 5052	0	28	13	25
SAE 5052	H32	33	28	12
SAE 6061	0	18	8	25
SAE 6061	T6, T651	45	40	12
SAE 7075	T6, T651	83	73	11
SAE 7178	T6, T651	88	78	10

Wrought Alloys

The first range of aluminum alloys to be used in high-performance structures had copper as their major alloying element. These "duralumin" alloys date back to 1909, and are now identified as "2000" series. Of these, 2024 is surely the most common. Its nominal composition is 4.4 percent copper, 1.5 percent magnesium, and 0.6 percent manganese, the remainder being pure aluminum. Depending on the details of heat treatment, 2024 alloy can provide up to 70ksi UTS, and yield strengths up to 66ksi.

Another common family of heat-treatable aluminum alloys—the "7000" series—have zinc as their principal alloying element. The nominal make-up of 7075 alloy, widely used in aircraft and race cars, includes 5.6 percent zinc, 2.5 percent magnesium, 1.6 percent copper, and 0.23 percent chromium. 7075 is even stronger than 2024, with a UTS in the T6 temper of 86ksi, and a yield strength of 73ksi.

With strength values like this, and specific stiffness that matches high-performance steels—and thus makes aluminum, with one-third the density of steel, clearly preferable for thin sections in compression or shear—it is reasonable to ask why it does not replace steel in virtually every race car application.

First is the matter of fatigue (see chapter 3). Aluminum fatigues significantly worse than titanium, composites, or steel. While steel, for example, can endure virtually forever under a cyclic stress corresponding to about half the static strength, most wrought aluminum alloys are down to around one-third of their initial strength after, say, 10 million cycles. Paradoxically, while the fatigue resistance of steels tends to improve as their ultimate strength increases, stronger grades of aluminum show little improvement in fatigue properties over many initially weaker ones. In some cases, an aluminum alloy with ho-hum performance in static tests will, over the long haul, exceed the strength of a "better" grade in fatigue tests.

Another catch is the problem of joining parts. The highest strength aluminum alloys, like 7075 and 2024, are essentially not weldable, which is why aircraft are riveted together rather than welded. (Adhesive bonding is a possible alterna-

As its long, successful history of use in aircraft, race cars, and other high-performance applications proves, aluminum's drawbacks are very frequently outweighed by its inherent advantage—lower density than steel at equivalent specific strength and stiffness.

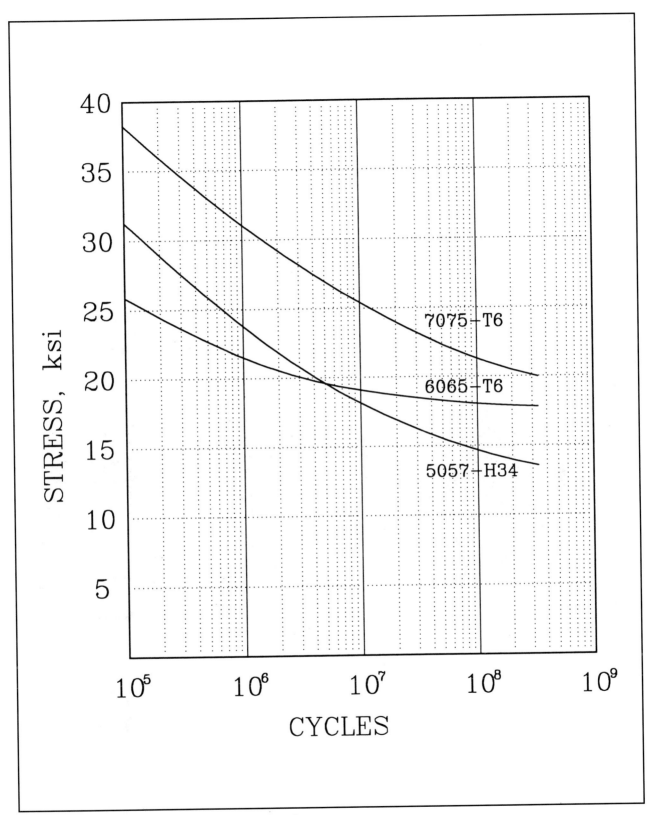

Some high-strength aluminum alloys fatigue worse than
some lower strength ones. In this case, if the design life of
the part was more than about five million cycles, the "weaker" 6065 alloy would be preferable to the "stronger" 5057.

To take advantage of high-strength aluminum alloys, which cannot be welded, this ultra-economy car is bolted together throughout. *Forbes Aird*

tive—they've been gluing airplanes together for years. Successful adhesive bonding of aluminum, though, requires close attention to the details of joint design, special chemical preparation of the surfaces, and purpose-designed epoxy adhesives. Worst, the soundness of such connections cannot be confirmed by visual inspection.)

Of the heat-treatable alloys, only the "6000" series are weldable. Of these, the most commonly used is 6061, which is alloyed with 1 percent magnesium, 0.6 percent silicon, 0.28 percent copper, and 0.2 percent chromium. In the softer tempers, it can readily be bent or otherwise formed to shape, while in the "hard" T6 temper, it offers ultimate and yield strengths of 45 and 40ksi, respectively. Strength "as-welded," however, falls to just 24 and 20ksi, respectively. And welded aluminum fatigues even worse than the same alloy

Despite the labor involved, and the problems of fretting corrosion and having to re-work them from time to time as they become "wiggly," road race cars and aircraft are also riveted (or bonded) together. *Forbes Aird*

Riveting is prohibitively expensive for mass production, so Acura uses weldable grades of aluminum for the structure on their NSX sports car. Despite the weight penalty this entails, compared to a riveted structure in a stronger alloy, Acura claims a 40 percent weight reduction compared to a steel body-frame. *Honda Canada*

before welding. (While virtually all of the original mechanical properties are restored if the finished assembly is heat treated afterwards, this is not usually practical.)

While aluminum generally has good corrosion resistance, alloys containing copper or zinc—ironically, the strongest ones—do less well than the others.

Ironically, some lower strength alloys that are not heat-treatable—but are weldable—provide higher "as-welded" strengths. Two relatively new alloys, 5083 and 5456, offer about 40ksi UTS and 25ksi yield in the "as welded" condition, compared to about 47 and 37ksi, respectively, in a work-hardened condition before welding. (All "5000" series alloys have magnesium as their principal alloying element, and contain no silicon or copper.)

Then there is the matter of corrosion (see chapter 3). While aluminum generally has good corrosion resistance, alloys containing copper or zinc—ironically, the strongest ones—do less well than the others. Further, because any two dissimilar metals in close contact try to behave like a battery, problems can arise when aluminum is bolted or riveted to a steel part, or when two aluminum parts are joined with steel fasteners. A thin surface layer of a more "anodic" alloy greatly improves the corrosion resistance of high-strength aluminum sheet, plate, or bar, and such "clad" products are readily available. There is a slight reduction in strength, in proportion to the thickness of the cladding—usually about 5 percent.

All these factors—fatigue, joining, corrosion—require consideration when using aluminum. In many applications, however, they can be conquered by suitable alloy selection, cladding, or other corrosion protection where necessary, and appropriate design. As its long, successful history of use in aircraft, race cars, and other high-performance applications proves, aluminum's drawbacks are very frequently outweighed by its inherent advantage—lower density than steel at equivalent specific strength and stiffness.

And there are situations where the low density itself is a deciding factor—sometimes the size of a part is established by considerations other than strength or stiffness. An engine block, for example, has to be big enough to contain all those 4in diameter holes, and there are many places in a casting where how thin the walls can be made is not limited by their strength, but simply by how thin a section it is possible to cast reliably, whatever the material.

While steel, for example, can endure virtually forever under a cyclic stress corresponding to about half the static strength, most wrought aluminum alloys are down to around one-third of their initial strength after, say, 10 million cycles.

Casting Alloys

Aluminum is readily castable by every means known to foundry men, and there are over 100 registered aluminum casting alloys, counting minor variations. Many of these are intended for mass-production casting in metal molds, either as "permanent mold castings" (the molten metal is simply poured in under gravity) or as "die castings" (forced into the mold under considerable pressure), and are not dealt with here. Of the rest, there are perhaps half a dozen alloys appro-

Aluminum lends itself to casting in very limited production numbers. Only a few dozen of the sand cast rear uprights of this small formula car were ever made. Even "one-offs" are sometimes economically practical. *Forbes Aird*

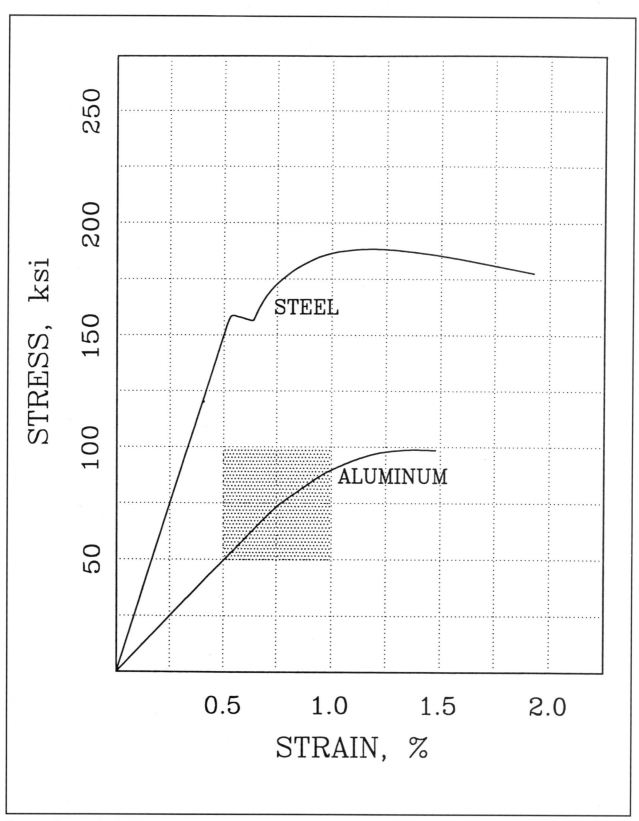

Compared to steel, aluminum does not exhibit the same
sharp "kink" in the stress:strain curve at the yield point; it just
gradually fades from "elastic" to "plastic."

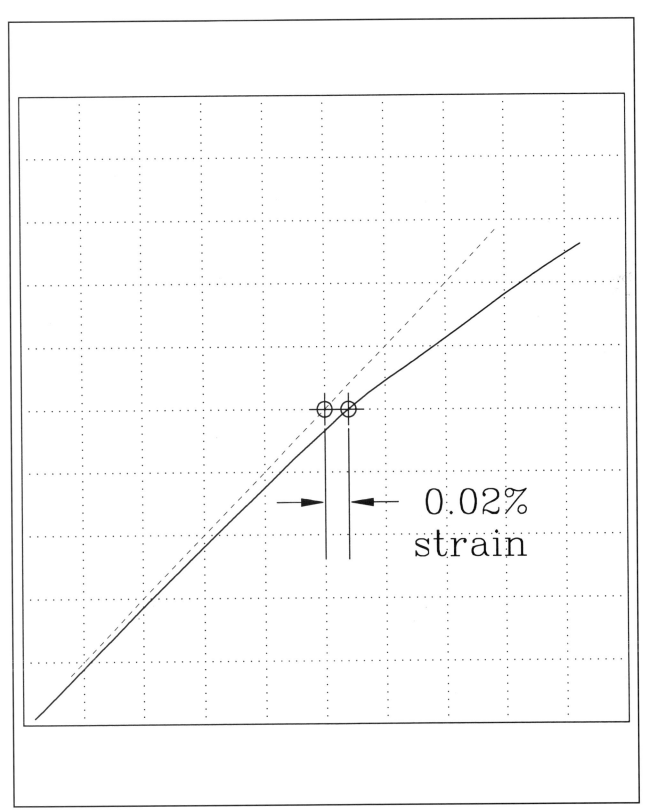

In the absence of a clear marker, the yield point for aluminum is defined to be the stress level at which the stress:strain curve has departed from a straight line by 0.02 percent strain.

priate for sand casting—a procedure well suited to limited production. It is sometimes surprisingly economical to get special parts sand cast in very small quantities.

For a handful of custom race car parts, the alloy selected is often 356 or A356, containing about 7 percent silicon. (The "A" prefix indicates a slight modification of the basic chemical composition.) Heat-treated to the T6 condition A356 typically gives 37ksi UTS, 27ksi yield, while 356-T6 shows about 33 and 24ksi. Elongations are about 5 percent and 3.5 percent, respectively.

This is perhaps the best combination of strength and ductility available from traditional sand cast aluminum—the chemically similar 355 alloy, for comparison, can be heat treated to give as much as 36ksi yield strength, but elongation is down to 0.5 percent at this strength level. Similarly, the copper/nickel/magnesium alloy 242 has equal UTS and about 10 percent better yield strength in the T4 temper than A356-T6, but ductility is inferior—elongation is limited to 1 percent.

While this accounts for the popularity and widespread use of A356, there are a couple of comparatively new alloys (356 goes back decades) that promise a significant improvement in both strength and ductility, though they are both more difficult to cast. Alloy 771, containing about 7 percent zinc, can show as much as 52ksi UTS, 48ksi yield, and 3 percent elongation, in the T71 temper. In a much more impact-resistant T6 temper, it still returns 45ksi ultimate and 37ksi

> *It is sometimes surprisingly economical to get special parts sand cast in very small quantities.*

yield, with 7 percent elongation. 771-T6 also claims a design fatigue limit of 10ksi (presumably after 10,000,000 cycles), versus 5ksi for 356-T6.

Even more remarkable is A206, a copper/magnesium alloy. In the T7 temper, it has typical specs of 63ksi UTS, 50ksi yield, with 11.7 percent elongation! As much as 17 percent elongation is available in a more damage-tolerant T4 temper, which returns 51ksi ultimate, 36ksi yield.

Magnesium

With a density of 0.063–0.066lb/cu-in, magnesium is less than one-quarter as dense as steel, and 35 percent "lighter" than aluminum. Its modulus of elasticity, compared to those metals, is lower in almost exactly the same proportion. We have already argued that "when there are compression loads that tend to make a part buckle, and when the various candidate materials have the same specific stiffness and strength, a less dense material will always save weight." Following that logic, if aluminum is good, magnesium must be better.

Some people have thought so—every Volkswagen "Beetle" contained 40-45lb of the stuff, mostly in its engine and transmission cases, and the Porsche 911 uses over 100lb per car. The "Beetle" is gone, though, and the average magnesium content of US-built cars and trucks today is less than 2lb, partly because of the current assessment of the cost of magnesium versus the value of a pound saved.

Certainly there is a fair amount of magnesium used in race cars. Since Ted Halibrand introduced them in 1947, cast or forged "mag" wheels have become universal in most serious forms of racing. And other bulky, thin section parts—like bell-housings, cam covers, transmission cases, and hollow suspension uprights—are routinely cast in the same material. Some people

Some Cast Aluminum Alloys

Alloy	Heat Treatment	UTS ksi	Yield ksi	Elongation %
206.0	T4	51	36	7
206.0	T7	63	50	11.7
242	T4	37	30	1
242	T77	30	23	2
355.0	T6	35	25	2.5
355.0	T7	38	36	0.5
356.0	T51	25	20	2
356.0	T6	33	24	3.5
A356.0	T51	26	18	3
A356.0	T6	37	27	5
771.2	T6	45	37	7
771.2	T71	52	48	3

The Jaguar D-type sports car—which dominated the 24 hour race at Le Mans, France, in the mid-fifties—was perhaps the most successful ground vehicle application of magnesium structures. *Jaguar Canada*

have gone further. The D-type Jaguar sports car, which won the famous 24 hour race at Le Mans, France, in 1955, 1956, and 1957, was one of the earliest applications of "stressed skin" construction in racing. The D-type's structural tub was made entirely from riveted magnesium sheet.

The Halibrand Shrike, designed for the 1964 Indy 500, achieved a dry weight of just 1140lb partly by the extensive use of magnesium, including brake calipers, suspension uprights, trans-

Apart from a tubular sub-frame used to support the engine and transmission, the entire structure of the D-type was riveted magnesium sheet. If you imagine the hood removed from the second car on the production line, everything else you see was mag structure. *Jaguar Canada*

mission case, and, most notably, a stressed-skin monocoque comprising 0.064in thick skins over wrought magnesium, wrapped over cast bulkheads of the metal. In Formula 1, at about the same time, the Italian ATS team produced a magnesium F1 monocoque entirely cast in one piece! Not much later, in 1967, Dan Gurney substituted magnesium for the aluminum skins in his Formula 1 Eagle, saving something like 50lb in the process.

For castings, at least, the magnesium alloy used in most of the above instances was AZ91—the "A" stands for aluminum, the "Z" for zinc, while the digits "9" and "1" indicate the nominal percentages of each added to the base magnesium. AZ91 remains the most popular magnesium alloy for race car castings, with 40ksi UTS,

An odd property of magnesium is that its yield strength may be less in compression than in tension, sometimes significantly so, though this is more generally true of wrought alloys than of castings.

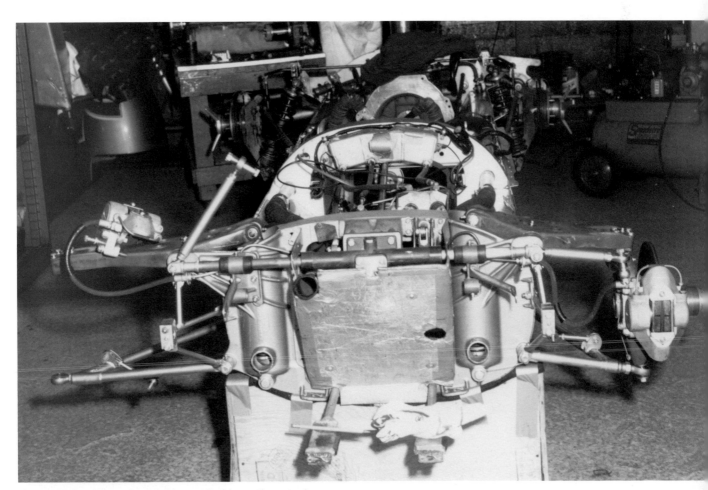

Magnesium's ability to be reliably cast in thin sections made the cast mag bulkheads on the 1964 Indy Halibrand Shrike practical. Wheels, suspension uprights, and many other pieces were likewise cast in the same metal. The chassis structure was completed by riveting sheet magnesium over the cast bulkheads. *Indy 500*

19ksi yield, and 5 percent elongation in the T6 temper. For engine crankcases, Porsche uses a German alloy corresponding to ZE41 (the "E" stands for "rare-earth metals"—a family of little known metals, including cerium and yttrium). Though it has a somewhat lower UTS than AZ91 at room temperature, at higher temperatures ZE41 exhibits better yield strength and improved resistance to "creep."

Magnesium is also available as wrought products—sheet, plate, and extrusions, such as tubing. While the use of structural magnesium tubing in race cars is rare, some Porsche race cars, including one of Roger Penske's 917 Can-Am entries, had their entire tubular frame welded from magnesium. In this case, the total frame weighed 100 lb, saving 33lb over the aluminum frame it replaced.

Perhaps the most commonly used wrought magnesium in race cars is AZ31 (another zinc/aluminum alloy), in the form of sheet and plate. Work hardened to the H24 condition, it offers 32ksi UTS, with 15 percent elongation. An odd property of magnesium is that its yield strength may be less in compression than in tension, sometimes significantly so, though this is more generally true of wrought alloys than of castings. AZ31, for example, provides a tensile yield

Despite the problems with welding, fatigue, and corrosion, a carefully designed and executed magnesium tube space frame can save weight, compared to either steel or aluminum. This seventies Porsche 917 Can Am car saved 33lb that way. *Porsche/VW Canada*

strength of 32ksi, but that value drops to 26ksi in compression. For sand cast AZ91E-T6, on the other hand, the corresponding figures are 21 and 19ksi.

Some Magnesium Alloys

Alloy	Form	Heat Treatment	UTS ksi	Yield ksi	Elonga- tion %
AZ31B	Wrought	H24	42	32	4
AZ31B	Wrought	H26	42	31	6
AZ91C	Cast	T4	40	14	7
AZ91C	Cast	T6	40	19	3
AZ91E	Cast	T6	40	21	6
HK31A	Wrought	H24	37	29	4
HK31A	Cast	T6	30	15	4
ZE41	cast	T5	30	20	–

There are several reasons why magnesium is not utilized more widely. While aluminum is used in huge quantities in many fields, much of the credit for the development of tough, high-strength alloys of that metal is due to the aircraft industry. And no small part of its widespread use in racing can be credited to that pioneering. But while magnesium has been and is used in aircraft, it has not been embraced with anything like the same enthusiasm shown for aluminum (which may explain why there are many fewer commercial magnesium alloys than is the case for aluminum), despite the apparent attractions of "mag."

Partly this apparent reluctance is because of concerns about corrosion. Magnesium is the most "anodic" of structural metals, which not only encourages galvanic corrosion when it is used in combination with other metals, but also ensures that it is the magnesium that will get "eaten" by the reaction. Worse, magnesium has had a reputation for stress corrosion cracking (see chapter 3).

Again, the skepticism of airframe makers may have something to do with the "scale effects" we touched on in chapter 2—it may be that the overall density of modern commercial aircraft is sufficiently great that, among metals, aluminum

makes the most sense. We must also remember that aircraft designers, though more weight conscious than most others, are also cost conscious; they too must attach a certain dollar value to each pound of weight reduction. (In helicopters, where a pound saved has higher value, much more magnesium is used.)

Note, too, that magnesium's density is in the same range as that of composites—less than those based on glass fibers, generally more than aramid-based RP, and about the same as CFRP. And the performance of composites outstrips magnesium by virtually all measures, except perhaps at temperatures between, say, 175-225 degrees F (at which point the plastic matrix of most composites starts to turn to mush) and 300-350 degrees F—the limit for "mag."

Finally, we can point to magnesium's relatively poor showing in terms of specific strength, the fact that wrought "mag" is more difficult to form, and concern about its flammability—though this last is often exaggerated.

The situation regarding magnesium may soon change. For one thing, the enormous quantity of electrical power needed to refine aluminum is pushing up the price of that metal, while there is a substantial overcapacity in the magnesium indus-

In response to concerns about corrosion, the magnesium industry has done considerable work over the last decade to demonstrate in practice what was known in theory in the forties—that magnesium's problems all stem from traces of heavier metals, especially copper, iron, and nickel, which had not been removed during refining.

Although differences in casting shrinkage (and the need, usually, for thicker sections) prevents direct substitution of mag for aluminum, the re-working of patterns is sufficiently straightforward that many racing parts are offered in either metal. Halibrand's quick-change rear end for sprint cars—here in mag form—is one example. *Halibrand*

try. This is shifting the relative prices of the two light metals at the same time that fuel economy demands are increasing the value of a pound saved, and industry experts claim that designs now being readied for production will triple the amount of magnesium in production cars. Note, too, that magnesium is generally easier to machine than aluminum—a significant cost factor.

In response to concerns about corrosion, the magnesium industry has done considerable work over the last decade to demonstrate in practice what was known in theory in the forties—that magnesium's problems all stem from traces of heavier metals, especially copper, iron, and nickel, which had not been removed during refining. New high-purity versions of familiar alloys (often identified by the letters "HP" or "UX" added after

The D-type Jaguar sports car, which won the famous 24 hour race at Le Mans, France, in 1955, 1956, and 1957, was one of the earliest applications of "stressed skin" construction in racing.

the basic spec number) show dramatic improvement in corrosion resistance. In one corrosion test, for example, the surface of "regular" AZ91 containing up to 140 parts-per-million (ppm) of iron corroded at the rate of about 0.312in per year! An identical surface of AZ91HP, which contains less than 50ppm of iron, showed less than 0.010in per year of surface loss.

The situation regarding magnesium may soon change. For one thing, the enormous quantity of electrical power needed to refine aluminum is pushing up the price of that metal, while there is a substantial overcapacity in the magnesium industry.

Titanium

Iron and steel have been around for centuries, aluminum and magnesium have been in use since before World War I, but structural applications of titanium are really quite new—newer even than composites. With a density of about 0.165lb/cu-in, titanium "weighs" 65 percent more than aluminum, and more than two and a half times as much as magnesium. It is about 45 percent lighter than steel, however, and since high-strength titanium alloys match or exceed the ultimate and yield strengths of all but the very strongest super alloy steels, they offer unmatched specific strengths. Titanium also resists fatigue exceptionally well (even better than steel), has exceptional corrosion resistance, and retains its strength at high temperatures.

These qualities have made titanium very popular in parts of the aviation and aerospace industry—nearly 10 percent of the airframe weight of the Boeing 747, for example, is made from the stuff, and in the YF 12 military aircraft (the fighter version of the SR 71 "Blackbird" spy plane) the proportion was 93 percent! An aerospace bolt of titanium alloyed with 1 percent aluminum, 8 percent vanadium, and 5 percent iron claims not only the highest specific strength of any available threaded fastener (230ksi UTS, typically, with 200ksi guaranteed, from a material with a density of 0.165lb/cu-in—a steel fastener would have to achieve 350ksi to match that), but also the highest fatigue strength-to-weight ratio. (A fatigue endurance limit of 56ksi was reported for the same bolt.)

Racers, too, have made use of the stuff. The "Mag-Ti" Gurney Eagle that saved weight through substituting magnesium for aluminum also gained lightness by replacing many steel parts (all suspension links, the front suspension rocker arms, and the entire exhaust system, among others) with titanium. Various Porsche race cars have used titanium for springs, axles, hubs, differential cages, disk brake "top-hats," and connecting rods, as well as smaller parts, including various bolts and screws. (In one particularly interesting application, ultra high-strength steel rod bolts were secured with titanium nuts, taking advantage of that metal's greater elasticity to relieve the high local stresses where the first bolt thread engaged in the nut.) Connecting rods in all current Formula 1 engines, and in the production Acura NSX sports car are made from titanium, and it is widely used for valves and valve spring retainers in many classes of racing.

On the other hand, titanium has some formidable drawbacks. First of all, it is expensive. It is also difficult to weld, and it is horribly difficult to cast—molten titanium corrodes most materials, so investment casting or rammed graphite molds are needed. While commercially pure titanium is claimed to be no more difficult to machine than stainless steel, high-strength alloys can be trickier, because of their inclination to stick to cutting tools, and this same "galling" tendency means that titanium cannot be used directly where there is relative movement between parts.

Another notable characteristic of titanium, its low thermal conductivity—about one-third that of steel, can be either a problem or a solution, depending on the situation. When Porsche first used it for front wheel spindles, they found that the wheel bearings ran much hotter than on a steel axle—up from 356 degrees F to 752 de-

In what is claimed as the first use of titanium in a production vehicle, the Acura NSX uses titanium connecting rods. A 30 pervent weight saving is claimed over a steel rod. The titanium alloy is produced using machining scraps left over from aerospace applications. This, and the development of a special coating technique to prevent galling, makes the weight saving cost effective. *Honda Canada*

Some Titanium Alloys

Alloy	Heat Treatment	UTS ksi	Yield ksi	Elongation %
6Al-4V	Annealed	144	134	4
6Al-4V	Solution Treated & Aged	170	159	10
4Al-4Sn-4Mo-0.5Si	HT	190	174	–
1Al-8V-5Fe	HT	230	210	13.8
5Al-2Sn	HT	115	110	–

grees F. (This problem was eventually side-stepped by the use of special high-temperature lubricants.) On the other hand, this same insulating property, and titanium's good retention of strength at elevated temperatures, makes it ideal for firewall material and as an insulating spacer between pads and pistons in disk brakes, to reduce heat input to the brake fluid. Wilwood, for one, has made the caliper pistons themselves from titanium, for the same reason.

The welding problem can be overcome by carrying out all welding in a special chamber purged of air and filled with inert gas. (Some titanium suppliers insist that completely satisfactory welds can be achieved with *some* alloys with simple inert gas shielding, similar to the technique used for TIG welding of aluminum.) Machining difficulties can be minimized by carefully tailored cutting tool angles and other techniques, including vast quantities of special water-based cutting lubricants, and almost completely eliminated if parts are chilled to sub-zero temperatures during machining operations.

In finished parts that are exposed to rubbing contact, like connecting rod thrust faces, the galling problem can be licked by flame or plasma sprayed coatings, typically of molybdenum or chromium nitride, as used by Honda on the NSX rods; by hard chrome plating; by a nitriding process involving a two-hour soak at 1500 degrees F in a cyan bath (this was Porsche's technique); or by oxygen deposition (as developed by Honda for con rods for a limited production motorcycle)—essentially heating in air at about 1300 degrees. All these techniques leave a surface suitable for use as a bearing.

Though no titanium alloy can really be regarded as "traditional," one alloy has proved to be the workhorse in aerospace and racing applications. Titanium alloyed with 6 percent aluminum

and 4 percent iron can be heat treated to a UTS of about 155ksi, with 145ksi yield strength. Titanium's inherently higher ratio of yield to ultimate strength means that, even with this small spread between yield and rupture, at this strength level the 6Al-4Fe alloy still provides 6 percent—8 percent elongation and 15–20 percent reduction of area. 6Al-4Fe (Fe is the chemical symbol for iron) is used in many aerospace fasteners, is the most common material for titanium intake valves, and was the alloy used by Porsche for many parts, including con rods, though it was later replaced in that application by a 7Al-4Mo (7 percent aluminum and 4 percent molybdenum) alloy that accepts heat treatment better and so offers slightly higher strength in thick sections. The alloy used by Honda for the NSX rods was 3Al-2Fe. (Note that both the alloying elements and their ration is the same as for 6Al-4Fe.)

For race engine exhaust valves, an alloy with better resistance to creep at high temperatures is needed. 6Al-2Sn-4Zr-2Mo-0.08Si and, more recently 6Al-2Sn-4.0Zr-0.4Mo-0.4Si have been used for this application.

The basis for the "Blackbird" was an alloy containing 13 percent vanadium, 11 percent chromium, and 3 percent aluminum—about the strongest titanium alloy available, with a UTS above 213ksi, and one which retains its strength exceptionally well at high temperatures. Titanium, however, exists in two basic forms, depending on how the crystals making up the metal are arranged, a layout that is essentially stable at room temperature but which changes at high temperatures. The so-called "Alpha" form includes commercially pure titanium; the more "common" structural alloys are a mixture of Alpha and "Beta" phases. The 13V-11Cr-3Al alloy, however, is entirely Beta, which makes it stable at high temperatures (which is why it was chosen

Another notable characteristic of titanium, its low thermal conductivity—about one-third that of steel—can be either a problem or a solution, depending on the situation.

for the "Blackbird"), but which has the odd effect of reducing its Young's modulus—Beta titanium has an elastic modulus of 15 msi, compared with 16.5 for the Alpha and Alpha-Beta forms.

Despite its inherently high price (for one thing, other relatively expensive metals like sodium or magnesium have to be sacrificed in its production), no small part of titanium's reputation for flabbergasting cost is attributable to the inspection and certification procedures demanded of all aerospace materials, and the very strict limits on impurities in the raw material. There is some prospect for a significant reduction in the price premium for titanium parts, as Honda's efforts have demonstrated. By processing the machining scrap from aerospace applications through a plasma furnace, it has been possible to produce connecting rods with aerospace-quality strength and reliability for less than six times the price of a comparable steel part (specifically, a connecting rod). Using virgin stock supplied to aerospace specifications, the titanium part would be more than thirty-seven times as expensive as steel!

Some Light Metals Compared

Alloy	UTS ksi	Yield ksi	E Msi	D lb/cu in	UTS/D	E/D
7075 Aluminum	83	73	10.4	0.10	760	104
AZ80A Magnesium	55	38	6.5	0.066	833	99
6Al-4V Titanium	170	159	16	0.165	1030	97
Beryllium/38% Alum	60	29		0.076		384
Beryllium	80	45	42	0.067	1197	628
Magnesium/14%Li/1.25%Al	15	6.2		0.049		128
"Lital B"	81.2	75	11.6	0.092	881	126

Chapter 6

Specialty Metals

High-Density Metals
Lead

For all the preoccupation of racers with "light" materials (i.e. ones of low density), there are times when what is needed is something "heavy." For ballasting of chassis, either to meet a minimum weight limit or to permit a more extreme weight offset to one side, lead is often used. With a density of 0.412lb/cu-in, it is about 40 percent more dense than steel or iron.

CAUTION: While fairly harmless in solid form unless ingested, both metallic lead and the oxide which readily forms on the metal's surface are nevertheless toxic materials. High concentrations can cause central nervous system disease. Skin contaminated by contact should be thoroughly washed off with hot soapy water. Inhaling the vapors formed when melting lead can cause serious health effects, including lung cancer.

Lead is also a very soft and weak material, with a melting point below some internal engine temperatures (pure lead melts at 621 degrees F, and even small amounts of alloying materials, like tin, can further reduce the melting point). For engine balancing, then, a stronger material with better temperature resistance is required.

Tungsten and Tungsten Alloys

Re-balancing the internal parts of a race-modified engine sometimes presents a challenge; while it is generally easy to remove surplus weight, adding it can be difficult, especially when space is limited. A widely accepted solution is to insert "slugs" of a high-density material.

Density of Some "Heavy" Materials

Material	Density lb/cu in
Steel/iron	0.30
Lead	0.41
Mercury	0.49
HD17	0.61
HD18.5	0.67
Uranium	0.68
Tungsten	0.70

While iron is usually regarded as "heavy," there are several metals of considerably greater density. Some, like osmium, iridium, and platinum, are nearly twice as dense as lead—nearly three times as dense as steel. Unfortunately, these are rare and precious substances. Another candidate, depleted uranium, is available virtually for free—no one who has any wants it! Surprisingly, it is entirely legal to acquire, but so much paperwork is involved that is hardly seems worth the trouble. Besides, the surface of uranium oxidizes rapidly.

A practical material for this application is tungsten (approximately two-and-a-half times as dense as iron or steel), and the popular "heavy metals" used for race engine crankshaft balancing are all based on tungsten, alloyed with small amounts of nickel plus iron or copper. The alloying metals reduce the cost of the sintering process used to produce the material, and ease machining

Re-balancing a modified engine often requires that weight be added where there is little room available. One solution is to insert "slugs" of a tungsten based alloy, having two-and-a-half time the density of steel or iron. *Dave Emanuel*

of the finished product. ("Mallory metal" is widely used as a generic name for this product. Although Mallory Metallurgical Co. and General Electric both produced this material at one time, neither presently does so, and Mallory strongly objects to the use of its name in this connection.)

Heavy metal is commercially available in various grades which differ slightly in density, cost, and machinability. Generally, more tungsten increases the density, at the expense of machining qualities. The range, though, is not great, with a

Most North American aftermarket engine bearings intended for racing use a lining of copper-lead alloy.

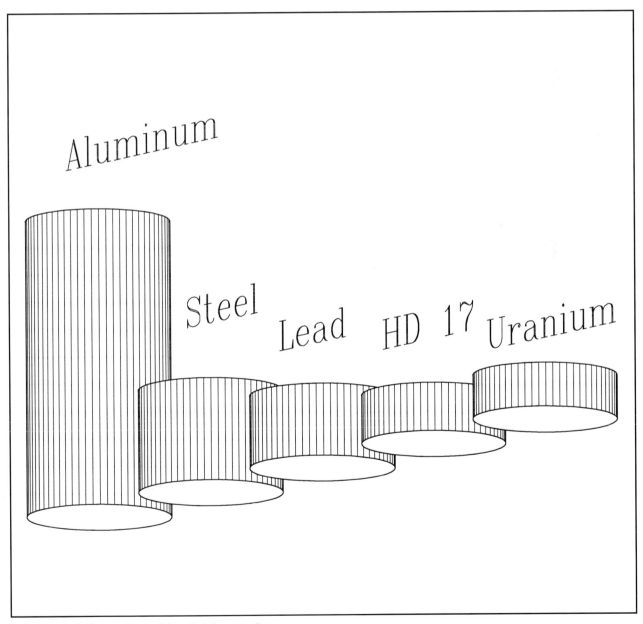

All these slugs of varied materials weigh the same!

> *Many aircraft engines in World War II used an intermediate layer of silver, which is very much stronger than copper-lead, has unequaled heat transfer capability, and excellent corrosion resistance.*

90 percent tungsten version weighing in at 0.614lb/cu-in, while 97 percent tungsten content yields 0.668lb/cu-in. It is usually supplied in the form of bar stock, often centerless ground to a precise finished diameter intended for a press-fit into a pre-drilled hole.

It is vital, however, that the press fit not be depended upon to retain the inserted plug, and that the plug be inserted parallel to the axis of rotation of the shaft—the coefficient of thermal expansion of this material is just 3.3 parts per million per degree F, about half that of iron. With heating, therefore, the hole would grow larger at about twice the rate of the tungsten alloy plug, so a one "thou" interference fit at room temperature would dwindle to zero when both materials reach about 400 degrees F. Heavy metal inserts are usually secured with a bead of weld.

Mercury

Mercury, the only metal that is liquid at room temperature, has a density of 0.489lb/cu-in. To circumvent rules in certain classes of racing that limit the offset of chassis weight, some people have used mercury as ballast material. The chassis is legal when weighed, but after opening a hidden valve, the mercury is allowed to flow through the frame tubes from the "wrong" to the "correct" side, shifting the weight distribution. While ingenious, this practice is not only unsporting, it is dangerous.

CAUTION: Mercury is toxic. Not only is it dangerous if ingested, it has sufficient vapor pressure that fumes are present even at room temperature, and inhalation of those fumes can be acutely hazardous. Permanent central nervous system damage can result from short-term exposure to high concentrations of mercury vapor, or longer term exposure to low concentrations. Also, the structural properties of aluminum are severely deteriorated by contact with mercury; its shipment by air is prohibited for that reason.

Bearing Metals

In an operating engine, the main and rod bearings do not actually contact the crankshaft but, rather, are floating on a thin wedge of oil. Nevertheless there are times, such as when starting, when something close to true metal-to-metal contact can occur. The first requirement for plain metal bearings, then, is that the bearing material must be "compatible" with the shaft—it must be inherently slippery to resist seizing under these conditions. It must also be comparatively soft, to accommodate slight misalignment and to permit hard particles such as dirt to embed itself, and so prevent damage to the shaft. The bearing material must nevertheless be strong (i.e. hard) enough to resist compressive fatigue failure under the working loads.

For many decades, a variety of alloys called "babbit metal" were used for plain engine bearings. Babbit—consisting of various proportions of lead and/or tin, together with antimony and/or indium and/or copper—is very soft and certainly meets the requirement for embeddability and "compatibility," but is limited in its fatigue strength, especially in thick sections.

Modern multi-layer bearings provide greater fatigue strength by adding an intermediate layer, or "lining," of strong, fatigue-resistant metal immediately next to the steel shell, and reducing the babbit to an "overlay" typically less than .001in thick, often electroplated onto the intermediate layer.

Linings

While providing a mechanical support for the overlay, the intermediate layer must also possess a certain amount of embeddability, and should have good thermal conductivity, to help draw heat away from localized hot spots. Most North American aftermarket engine bearings intended for racing use a lining of copper-lead alloy. The exact composition does not vary much from one manufacturer to another, generally consisting of 75–85 percent copper, with the balance made up of lead.

The copper-lead lining can be bonded to the steel backing by casting, in which case the material is actually melted onto the steel. Alternatively, it may be applied by sintering, which involves applying metal powder onto the shell, then rolling it at a temperature high enough to melt the lead constituent, but not the copper. The cast product is claimed to be about 20 percent stronger.

As an alternative to copper lead, various aluminum alloys have been used for the intermediate layer, particularly in Europe and Japan, but also by some North American engine makers. While their corrosion resistance is superior to copper-lead, aluminum is usually considered inferior in fatigue resistance, though some alloys can be heat treated to give comparable results. One such material developed in Europe for turbocharged diesel engine bearings, in fact, appears superior to copper-lead in fatigue strength. The particular alloy contains about 12 percent silicon and is thus chemically similar to the low-expansion aluminum alloys used for pistons.

Note, too, that a Land Speed Record car of the sixties, built by the Summers brothers, ran aluminum rods directly against the crank journals, with no bearing insert of any type. And, while it is not applicable to racing, some North American passenger car engines run an aluminum-lead alloy directly against the shaft, with no overlay. Fuel dragsters, on the other hand, use only the babbit and dispense with the intermedi-

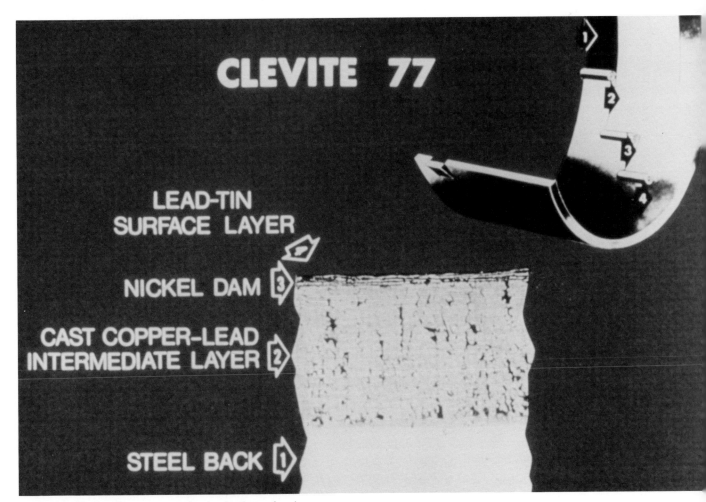

CLEVITE 77

LEAD-TIN SURFACE LAYER ▷
NICKEL DAM ③
CAST COPPER-LEAD INTERMEDIATE LAYER ②
STEEL BACK ①

Modern engine bearings are complex laminations of various metals, each chosen for a particular characteristic. Apart from the babbit surface, copper-lead lining and steel back (which, together, give "tri-metal" bearings their name), there is a micro-thin layer of nickel to prevent the tin in the babbit from leaching into the lining. *AE/Clevite*

> *The first requirement for plain metal bearings, then, is that the bearing material must be "compatible" with the shaft—it must be inherently slippery to resist seizing under these conditions.*

ate layer. They use a "micro-babbit" (i.e. the bearing material thickness is about 0.004-0.005in) applied directly to the steel shell, to provide enough conformability to accommodate the severe crankshaft distortions resulting from enormous cylinder pressures. Such bearings, though, have a service life of about five seconds!

Given adequate shaft stiffness and satisfactory oiling, the demands of most race engines are still within the capacity of overlaid copper-lead bearings, but if a stronger intermediate layer is needed in the future, it can be found in history. Many aircraft engines in World War II used an intermediate layer of silver, which is very much stronger than copper-lead, has unequaled heat transfer capability, and excellent corrosion resistance. It is also, of course, very expensive.

Overlay Metals

Most overlays are lead-tin alloys—a typical composition is 87 percent lead, 10 percent tin, and 3 percent copper. Generally, increasing the proportion of tin raises the fatigue strength and corrosion resistance, but reduces embeddability, requiring a harder journal. To prevent the tin content from migrating out of the babbit into the intermediate layer, an extremely thin (0.00005in) "dam" of nickel is set between the lining and the babbit. Partly to avoid the need for this extra layer, some car makers, particularly in Europe and Japan, use another soft, slippery metal, indium, together with or instead of the tin. Generally, lead-tin alloys are slightly stronger than lead-indium, but somewhat less soft and conformable. Indium is also more expensive than tin.

Expansion-Control Metals

When different materials are used in the same component, a mismatch in their thermal expansion can create unexpected stresses in one or both parts. For parts that operate at room temperature, this is seldom a problem, but some engine and brake components experience a range of temperatures that may extend from below zero to a thousand degrees or more, which will result in a considerable change in size. Even if both materials have sufficient strength at either temperature extreme, the physical "fight" between them may overstress one or the other.

Invar

This alloy of 64 percent iron and 36 percent nickel has the unusual and valuable property of a thermal expansion coefficient which is essentially zero over a wide range of temperatures—it does not expand or contract when heated or cooled. Its density and elastic modulus are about the same as those of other steel alloys. While this material, "Invar," is mostly used in scientific and optical instruments, it is also used as an insert in the pistons of some diesel engines, to control piston expansion. While no race car uses are known, applications involving ceramics—which have a very low expansion coefficient compared to most metals—may call for this property in adjacent metal parts.

Dilavar

There are places where what is wanted is not less expansion with heat, but more. Steel studs joining aluminum or magnesium engine parts, for example, "grow" considerably less than the surrounding light metal. Porsche encountered problems resulting from this effect with the air-cooled engines in some of some of their sports racing cars in the seventies. The long studs which secured the aluminum heads and cylinder barrels to the magnesium crankcase would either be too short when cold, or too long when hot. If installed slack, early fatigue failure would result; if appropriately tightened when cold, the consequence was distorted parts or stripped threads once running temperature was reached.

Neither steel (with an expansion coefficient of about 6.5 parts per million per degree F) nor titanium (4.5 ppm per degree) was suitable to match the expansion of the light metal parts, at roughly 13 to 14 ppm per degree. The solution was found in a range of steel alloys, trademarked "Dilavar" by Thyssen Edelstahlwerke. Dilavar Ni13, containing about 13 percent nickel, 5-6 percent manganese, 3-4 percent chromium, and dashes of other alloying metals, exhibits a thermal coefficient of about 11.3 ppm—closely matching aluminum and magnesium. It has about the same density and elastic modulus as other steels and, in the temper chosen by Porsche, it delivered a UTS up to 180ksi and a yield strength of at least 116ksi, with more than 12 percent elongation.

Copper-chromium for Brake Rotors

Heat enters race car brakes far faster than it can get out, so the brakes have to store the heat created while braking, mostly in the mass of the rotor material, then dissipate that heat to the surrounding air later. The preliminary selection of material for a race car brake rotor, then, is usually based on two benchmarks—the quantity of heat it takes to raise the temperature of a pound of the material by one degree (called "specific heat"), and how hot the material can be allowed to get before it begins to wilt.

For example, in rising one degree F in temperature, a pound of aluminum will soak up about 0.2 BTUs (British Thermal Units) of heat—twice as much as a pound of iron. So, for the same weight, an aluminum rotor will run cooler than an iron rotor. But aluminum loses strength rapidly as its temperature rises, so the total amount of heat the aluminum rotor can store and still hang together will actually be less than the iron one.

---

*The rotors tested by
and for Ford—both iron and
copper-chromium—employed
a flame-sprayed hard facing,
consisting of copper with about
3 percent silicon carbide,
to provide a harder
rubbing surface.*

Thermal Expansion of Some Materials

Material	Expansion coefficient (parts-per-million per degree F)
Structural Metals:	
Cast Iron	6.5
Mild Steel	6.33
Nickel Superalloys	7.5 to 8
Aluminum Alloys	10.7 to 13
Magnesium Alloys	14
6Al-4V Titanium	4.6
Beryllium	6.7
"HD17"	3.3
Expansion Control Metals	
Invar	0
"Dilavar"	11.3
Ceramics, etc.	
Al/60% SiC	4.7 to 6
Silicon Carbide	1.5 to 2.2
Aluminum Oxide	4.4
Carbon-carbon	-0.6 to +1.2
Fibers	
E-glass Fiber	2.8
"S-glass" Fiber	3.1
Aramid Fiber	-1.1
Carbon Fiber	-0.9 to +1.0

Some materials with exceptional values of both specific heat and hot strength (beryllium, carbon-carbon, and ceramic-reinforced aluminum) are discussed in chapter 11, while the properties of aluminum and iron are considered in chapters 5 and 4, respectively. There is one other material worthy of consideration for brake rotors, however—copper alloyed with 1 percent chromium; while sometimes called chromium-copper, we refer to it here as copper-chromium.

Compared with cast iron, copper-chromium has slightly greater density and closely comparable specific heat and service temperature limit, so by the criteria suggested above, it appears to have no particular advantage, while costing considerably more. Its appeal, however, lies in its extremely high thermal conductivity—twice that of aluminum, six times iron's—which helps to minimize temperature differences between various parts of the rotor. For the same *average* rotor temperature, then, the temperature at the area of pad contact will be lower, helping to reduce both pad wear and variations in friction coefficient through the working temperature range. At the same time, the temperature at the cooling vanes will be somewhat higher, helping to dissipate the stored heat.

These factors led to experiments with this material by Ford, when they ran into brake cooling problems during the development of their (eventual) LeMans winning 1966 GT Mark II cars. Porsche, too, tried copper-chromium discs, and won the 1970 Daytona 24-hour race using them, though they later found that cross drilled iron discs could be made lighter. Ford experienced teething troubles with the copper-chromium rotors, and never used them in a race. Nevertheless, their experience suggests that copper-chromium has advantages under certain circumstances, specifically for applications that require fairly high total heat storage capacity, but which suffer problems with varying friction coefficient because of wide swings in pad temperature. In some of these cases, pads with a friction coefficient that drops off less at high temperatures may present an unacceptable trade-off, because of insufficient friction at low temperatures.

Generally, more tungsten increases the density, at the expense of machining qualities.

The rotors tested by and for Ford—both iron *and* copper-chromium—employed a flame-sprayed hard facing called DiPac, consisting of copper containing about 3 percent silicon carbide, to provide a harder rubbing surface. Other experimenters have found that copper-chromium provides an acceptable rubbing surface for both discs and drums, even without a protective facing, though wear rates on the copper-chromium may be high when used in conjunction with a friction material having an abrasive character.

Comparison of Brake Rotor Materials

	Thermal Conductivity*	Specific Heat #	Density lb/cu in	Melting Point (deg F)
Cast iron	30	0.11	0.29	2750
Chromium copper	187	0.09	0.32	2000(?)
Aluminum	96	0.21	0.10	1220
Beryllium	63	0.23	0.067	2350
Carbon/carbon	7.4	0.17	0.06	4500

* Btu/hr/sq ft/deg F/ft (@ room temp.)
# Btu/lb/deg F (@ room temp.)

Chapter 7

Thermoplastics, Elastomers & Wood

Wood—The Material that Grows on Trees

It may seem bizarre to include reference to wood in a book on race car materials, but there are two points to be made. First, we have already noted that good quality wood is about as strong and stiff in relation to its weight as any other traditional material, and that its low density means that a wooden panel of a given weight will be more resistant to buckling than one of the same weight made from any common material that is more dense—and that covers all of them! Second, many successful race cars have been made with wooden construction. Apart from the brilliant

Bearing in mind the limiting factor in the design of stressed skin structures—that of buckling under compressive and shear loads, Douglas fir remains a viable material for this form of construction, by virtue of its very low density.

lightweight Special built in the thirties, and McLaren's Formula 1 "Metalite" chassis in 1965—both of which used wood as a core material for a structural "sandwich" (see chapter 10)—there is the work of Frank Costin, an ex-de Havilland Aircraft engineer.

Costin had designed and built various gliders with wooden structures, so he recognized the advantages of making both body and chassis from a material—plywood, probably Douglas fir—that was cheap, readily available, and easily manufactured by semi-skilled labor, using the simplest of equipment. The result was a body/chassis unit weighing just 145lb that nevertheless possessed a torsional stiffness of over 3,000ft-lb per degree. Eleven such cars were built, some of them very successful in sports car racing in the early sixties, before the business side of the venture got out of hand.

As unprocessed lumber, Douglas fir has a UTS of perhaps 20ksi and a Young's modulus of about 1.6msi, when measured parallel to the grain. Considered in relation to its density of about 0.2lb/cu-in, this puts Douglas fir in the same category as heat treated alloy steel and aluminum on a tensile-strength-to-weight basis, and only about 10 percent inferior to them in terms of stiffness-to-weight. The strength of wood in compression, however, is much less than in tension—perhaps two-thirds less. And, naturally, the properties measured across the grain are as little as one-quarter the lengthwise values.

Stacking thin layers of wood with the grain of alternate layers at right angles produces plywood which, predictably, is superior to the cross grain properties of the lumber, but is much less strong than unprocessed lumber tested along the grain. As a result of this averaging of properties, the specific tensile strength and stiffness of plywood drops dramatically, though the compression properties are less affected. Bearing in mind the limiting factor in the design of stressed skin structures—that of buckling under compressive and shear loads, Douglas fir remains a viable material for this form of construction, by virtue of its very low density.

One feature of wood is that while the absolute strength, stiffness, and density differ widely from one species to another, the *specific* strength and stiffness is very nearly constant. Both strength and stiffness, in other words, vary pretty much in proportion to the density. Excluding the use of balsa as a core material, the range of wood densities suitable for stressed skin construction does not vary much—rather less than the relative difference between magnesium and aluminum, for instance. Using aircraft practice as a guide, the spread is from birch (0.20lb/cu-in) to Sitka spruce (0.16lb/cu-in).

Sitka spruce was used for wing spars in aircraft for many decades, because of its low density and the fact that it could be obtained as very

One of the most successful aircraft of World War II, the de Havilland Mosquito, was built entirely from wood. Here, workers apply a thick layer of end-grain balsa to the inside of a fuselage half formed from plywood. The balsa will later be covered by a second skin of plywood, forming an early and highly effective "sandwich" structure. *Bombardier Regional Aircraft*

long, continuous, straight-grained timbers. Sitka, however, grows only in a very narrow strip of land in the Northwest, along the Pacific coast, and recent demand for the attractive, white wood (particularly from Japanese guitar makers, of all things!) means that the supply of aircraft grade Sitka is now extremely limited. Douglas fir is regarded as a plausible substitute, with due allowance for its slightly greater strength and density. Birch, even more dense, is still widely used for the making of aircraft grade plywood.

For his next wooden chassis racer—the Protos Formula 2 car in 1967—Costin in effect made his own birch plywood, to take best advantage of the directional properties of his raw material. Birch strips just .138in thick and a hand-span wide were laid up at an oblique angle over a series of pre-formed plywood bulkheads, then a second layer glued over top at an opposite oblique angle. The entire skin weighed just 49lb. (This technique also helps ensure that the hidden layers are free of flaws. While not a problem with true aircraft grade ply, commercial grades may contain invisible faults. This is reflected in the published "allowable unit stresses" for commercial Douglas fir, which are less than a quarter of the theoretical strength calculated on the assumption of perfect plies.)

The adhesives used in the aircraft and marine industries are completely water resistant, and modern treatment can completely eliminate the problems of rot and fungus, so while wood might seem an eccentric choice of material for race car construction, these examples should demonstrate that it is entirely practical and that it would be wrong to reject any structural material on the grounds that it is "old-fashioned." Besides, the stuff grows on trees!

Thermoplastics

In its most general sense, the word "plastic" covers everything from printer's ink to foam rubber, but we can narrow the field a bit by limiting our discussion to "engineering" or "structural" plastics. They, in turn, can be divided into "thermosets" and "thermoplastics." Thermosets pass from liquid to solid by undergoing a chemical change; once solidified, they cannot be re-melted. Some thermosets of interest to racers are discussed in chapter 9. Then there are thermoplastics—the kind that can pass from solid to liquid and back again by heating and cooling.

There are thousands, perhaps tens of thou-

Properties of Some Thermoplastics

Material	UTS ksi	Elongation %	E msi	D lb/cu in
6-6 Nylon	11.8	60	0.41	0.041
Acetal Homopolymer	13.6	25	0.52	0.052
Acetal Copolymer	8.8	65	0.41	0.051
Cast Acrylic	8	4.5	0.47	0.043
Polycarbonate	9.75	80	0.24	0.043

sands, of thermoplastics, most sold as molding powders which are used to produce an endless variety of plastic goods by injection molding. To use them, you need an injection molding machine in which the powder is heated, then the molten plastic squeezed into a die cavity—plus, you need the dies. Otherwise, the field of interest is limited to thermoplastics that are available in the form of raw stock—sheets, rods, tubes etc.

Despite their light weight, thermoplastics lack both strength and stiffness compared to other engineering materials, and so are of limited value and interest where low weight and high performance takes precedence over low cost. Nevertheless, there are a few places where the properties of a particular plastic are either desirable or essential—the only alternative to a brittle, low-stiffness plastic windshield is a brittle, low-stiffness glass windshield, which is much heavier to boot.

Nylon

Nylon was the first engineering thermoplastic, and remains about the most common. It is tough, easily machinable, reasonably strong and, as plastics go, fairly stiff. One significant feature of thermoplastics is that their mechanical properties depend on the length of time over which the load is applied—a steadily applied load will cause the material to "creep." Their strength and stiffness also depends strongly on temperature; in fact, they behave more like viscous fluids than true solids, and the viscosity generally drops as the temperature rises.

Nylon is also "hygroscopic"—that is, it absorbs water from the atmosphere, and its strength drops as the water content increases. Still, at room temperature under *short term* loads, *dry* nylon exhibits a UTS of 11–12ksi and an elastic modulus of about 0.4msi. Taking into account its density of just 0.041lb/cu-in, that makes nylon's specific strength comparable to weak metals. Nylon also has good fatigue characteristics—its endurance limit (the load that can be applied and relaxed almost indefinitely without failure) is about 2.5–3ksi. Its specific stiffness, though, is only about one-tenth of the value shared in common by most metals and other structural materials.

Nylon has good chemical resistance, though it swells slightly in gasoline and is attacked by many acids, including sulphuric (battery) acid. It is used for small plastic screws, nuts, clips etc., and for various rubbing strips, due to its good abrasion resistance.

Acetal

Acetals are hard, crystalline plastics that were originally developed to replace metal (zinc, etc.) die castings. Acetal exists in two forms—"Delrin" (technically, an acetal "homopolymer") and "Celcon" (a "copolymer"). The homopolymer form is generally harder and stronger, but has less elongation.

Generally comparable to nylon in strength and stiffness, acetals have a very slippery surface and operate well without lubrication as bearings for moderate loads, at moderate temperatures. Accordingly, it is used for many of the "plastic" bushings found in control linkages etc. Acetal also has exceptional electrical properties, and it is widely used for ignition and electrical parts, such as distributor caps and ignition coils. Unlike nylon, acetal is essentially unaffected by moisture. It is resistant to most solvents and fuel components, but is attacked by certain strong acids.

Acetal also has exceptional electrical properties, and it is widely used for ignition and electrical parts, such as distributor caps and ignition coils.

Acrylic

Cast methyl methacrylate— "acrylic," for short—is transparent and is sold under a variety of brand names ("Plexiglas," "Lucite," "Perspex," "Rohaglas," etc.) as a lightweight glazing material. Comparable to nylon and acetal in stiffness, it is somewhat less strong than either and has a much lower impact strength. Its elongation is also very low (less than 5 percent, compared to about 60 percent for nylon), so it is extremely sensitive to stress concentrations. Still, acrylic has many times the impact strength of glass, and is one of only two practical alternatives to glass for situations that require optical transparency (though it is also available in colors, as well as in opaque form).

Cast acrylic can be cut and machined much like wood, though its notch sensitivity and its tendency to brittle fracture means that great care has to be taken to avoid cracking, especially when drilling holes. For the same reasons, and to avoid injuries, cut edges must be smoothed by scraping, sanding, or "flame polishing"—quickly passing the flame of a propane or oxy-acetylene torch over the rough edge. With skill and practice, this can produce a smooth, thickened bead, which also helps prevent future cracking. With bad luck, the entire piece will catch fire! Acrylic burns like wood.

Since it is a thermoplastic, acrylic can be shaped by first heating to a pliable state (at about 300 degrees, F), then draping over a form. A simple alternative method, which avoids both the need for a shaped form and the tendency of the soft, hot acrylic to pick up marks from any surface it contacts, is "free blowing." The sheet is heated, as before, but is then grasped around the edges in a frame which has a shaped opening exposing one side of the sheet. Applying compressed air then causes the acrylic to blow out of the opening, like a giant bubble. The shape of the perimeter of the bubble is established by the shape of the "port hole," and the height of the bubble by the amount of air pressure applied. This is an excellent way to produce windshield shapes for race cars and motorcycles, as well as aircraft canopies.

Polycarbonate

The alternative to cast acrylic as a lightweight replacement for glass is polycarbonate plastic, sold as "Lexan" by General Electric. Virtually identical to cast acrylic in density and strength, polycarbonate has but half of its stiffness, which makes it very willowy stuff indeed. It is also dramatically more expensive. Its great virtue is its tremendous impact resistance—30 times that of acrylic, 250 times that of glass. Polycarbonate is thus used widely for race car windshields, crash helmet face shields, etc.

It can be formed in much the same ways as acrylic, though it exhibits a much sharper melting point—the temperature difference between soft enough to drape and a runny mess is much narrower. Polycarbonate retains its strength and stiffness better than acrylic at high temperatures—while acrylic has a maximum service temperature of about 160 degrees F, polycarbonate will tolerate about 100 degrees more. Its transparency is unaffected by gasoline, motor oil, and many common solvents, though its strength will drop by up to 20 percent after long saturation by petroleum products.

Elastomers

Natural and synthetic rubber, and the many plastics that feel and behave like rubber, all make up the class of materials called elastomers. Almost without exception, they are used specifically where a material with a very low stiffness is desired, so it is not useful to discuss the strength and stiffness of these materials in the same terms as rigid structural materials. Apart from their use in tires, drive belts, and anti-vibration mountings, and (occasionally) their application for springs—none of which we will be discussing here, they are seen in race cars mostly as seals, gaskets, and flexible hoses for fluids. It is their resistance to these fluids that is thus of primary interest.

Nitrile

The most versatile general purpose elastomer for sealing and general fluid handling purposes is nitrile rubber, also called "BUNA-N" or "NBR." It possesses good resistance to most fluids found around race cars, including ethylene glycol (antifreeze), gasoline, methanol, and lubricating oils and greases, including silicone lubricants. Though it swells slightly after exposure to most petroleum products, this is often as much an asset as a drawback, as it helps to tighten up the seal. One of the few common applications for which nitrile is *not* suitable is for brake fluid.

Nitrile is widely used for seals, both the conventional lip type and O-rings, and its only real drawbacks are somewhat limited tolerance of both ultraviolet light and heat. Continuous exposure to temperatures over 250 degrees F should be avoided. It possesses good (though not outstanding) abrasion resistance, and so is suitable for both static and dynamic (moving) seal applications.

Fluorocarbon Rubber

Otherwise identified as "FPM," fluorocarbon rubber is also trademarked "Viton" by E.I. DuPont de Nemours Company. Like nitrile, it is resistant to petroleum products, but FPM is *not* resistant to methanol. Also like nitrile, it is not suitable for use with automotive brake fluid. It has the advantage over nitrile in that it is serviceable at continuous temperatures to 400 degrees F and will take higher temperatures (to 600 degrees F) for short periods.

BUNA S

The original synthetic rubber developed during World War II, BUNA S or "SBR" is one of few materials appropriate for automotive brake systems. Note, though, that its service temperature limit of 225 degrees F is well below the boiling point of racing brake fluid. It is severely attacked by petroleum products, swelling and softening greatly as a result.

Ethylene Propylene

Ethylene propylene or "EPM" is the other elastomer suitable for exposure to ester-based automotive brake fluid. It has the added benefit of tolerating higher temperatures than BUNA S, with common formulations serviceable to 300 degrees F, and some special formulations going even higher.

Like BUNA S, EPM should not be exposed to petroleum products like gasoline and lubricating oils, though it is suitable for use with methanol and is reasonably resistant to nitromethane.

Chapter 8

Plastic Composites— 1: Reinforcing Fibers

Introduction

A composite is simply a combination of materials which offers better performance than any one of the ingredients alone. Technically, it is difficult to draw a clear dividing line between composites and other combinations of materials, such as laminated safety glass, or reinforced concrete. For our purposes, though, the term composite means a material made by reinforcing an easily-worked but comparatively weak substance with fibers of something stronger. The strong fibers are called the reinforcement; the stuff surrounding the fibers is called the matrix. A great many combinations of materials can be used. Composites based on a matrix of metal or ceramic are mentioned in chapter 11, but their manufacture is a complex industrial process, requiring elaborate facilities and high temperatures. Materials that can be produced at room temperature, using simple, hand-held equipment is far more accessible.

Although this idea dates back to the ancient Egyptians, who included straw in clay bricks to reduce their brittleness, composite materials that can compete with metals on a strength-to-weight basis need high-performance reinforcing fibers, and matrices to match.

The great grand-daddy of modern composite materials was "fiber glass"—very fine filaments of glass embedded in a matrix of plastic resin. (The word "Fiberglas" is a registered trademark of Owens Corning; a more general term is "Glass Fiber Reinforced Plastic"—GFRP.)

It may seem puzzling that the combination of a brittle material, like glass, with a weak and soft material, like plastic resin, can combine to produce a strong and tough material like GFRP. The trick lies in the very fineness of the fibers and in the way they interact with the resin matrix.

In Part I we discussed how stress concentrations affect the strength of a piece of material, and how the focusing of stress caused by even a very small surface crack can initiate failure of the whole sample. While metals can yield to redistribute the stresses around the tip of the crack, non-yielding materials cannot. It is this inability of brittle materials to re-arrange themselves internally, so as to cope with surface flaws, that makes them weak.

Glass is a classic example of a material which, in bulk form, is both weak and brittle. Yet

For any given type of fiber, the factors of greatest significance are the volume fraction of fiber and the type of weave.

glass that is entirely free of any surface defects is, in fact, enormously strong. Anything that contacts that perfect surface, however, threatens to create surface cracks in the form of microscopic scratches. If a load is then applied to the glass, these cracks will rapidly progress through the material, and the glass will break.

For each material, though, there is some critical crack depth that can be tolerated, so if the material is drawn out into filaments thinner than

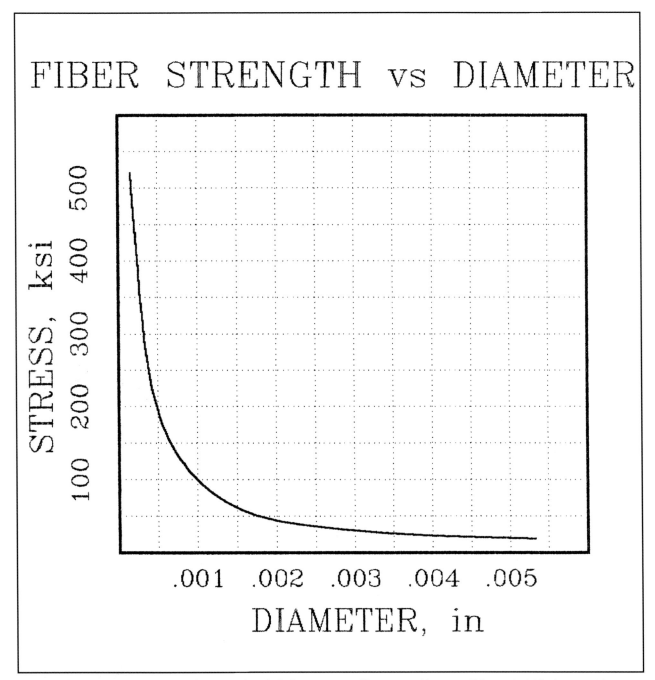

Many brittle materials are enormously strong, but their theoretical strength cannot be realized because they are unable to halt the advance of cracks arising from surface flaws. There is some lower limit, though, to the depth of crack that will progress. However, if the material is drawn out into very thin filaments, eventually a fineness is reached that is less than this "critical crack length," and the theoretical promise is achieved.

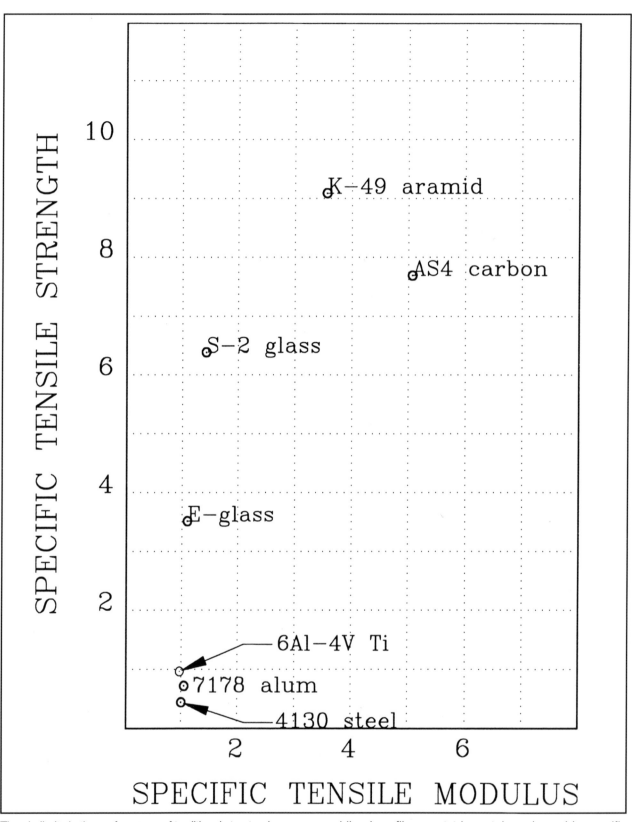

The similarity in the performance of traditional structural materials, and the exceptional properties of various fibers used in composite construction are apparent here. Note that while glass fibers outstrip metals and wood in specific strength, they show only a small edge in specific stiffness.

this critical dimension, the mere fact that a fiber is continuous pretty well guarantees that it is free of surface cracks. Also, because the fibers are so thin, anything that contacts them will tend to just push them out of the way, rather than scratch them, and even if a fiber does become damaged, it will simply break into two shorter pieces. In other words, a filament (of glass, in our example) is either completely intact or it is broken—unlike bulk glass, it can't appear OK yet still contain invisible flaws.

In the case of the other materials used for reinforcing fibers, comparative tests between fibers and bulk samples are not possible, simply because you cannot produce large chunks of the stuff. Nevertheless, the principle remains the same—fibers reflect the "real" properties of a material; bulk samples reflect its ability to cope with flaws.

Fibers for High-Performance Composites
E-glass

Glass is by far the most common kind of fiber used for reinforcement, and the vast majority of

Properties of Some FRP Laminates

Material	UTS (& Yield) ksi	E msi	D lb/cu in
E-glass Cloth /polyester	38	2.2	0.062
Uni-directional "S-2 Glass"/epoxy (57-63% fiber)	230-290	7.7-8.5	0.072
Aramid Cloth/ vinylester	52	3	0.042
Satin-weave Aramid cloth/epoxy	75	5	0.05
Uni-directional Aramid cloth/epoxy (65% fiber)	200	11	0.05
Satin-weave HS-CF/ Epoxy (55% fiber)	50	5.7	0.058
Satin-weave HM-CF/ Epoxy (55% fiber)	44	12.8	0.061
Uni-directional HS-CF/ Epoxy (65% fiber)	325	22	0.056
Uni-directional HM-CF/ Epoxy (65% fiber)	165	60	0.065

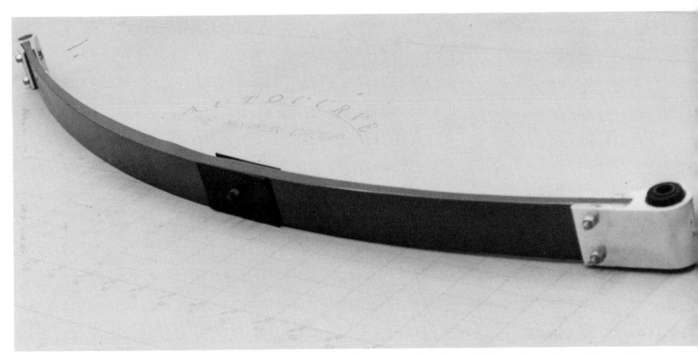

The great strength of glass fibers allows the construction of springs with a high factor of energy storage in relation to their weight. Apart from the GFRP springs used on Corvettes, racing aftermarket suppliers offer a wide variety of leaf springs in this material. *Flex-a-Form*

glass reinforcement is the same alumina borosilicate glass used for electrical insulation—hence the "E" designation. The fibers are produced by melting glass marbles in an electric furnace, then mechanically pulling the molten glass through tiny holes in a platinum bushing in the base of the furnace. The resulting continuous filaments of .000175–.000525in diameter are then gathered together into strands, and wound onto a reel.

Tests on single fibers of E-glass indicate tensile strengths of 500–550ksi (i.e. 500,000–550,000psi), and a tensile modulus (Young's modulus) of 10–10.5msi (10,000,000–10,500,000psi). When resin-impregnated strands are tested, rather than individual filaments, the strength numbers fall to 270–390ksi. Finished products using fiber glass reinforcement have even lower values—a quarter of the single filament's tensile strength (say 120–150ksi) and 60 percent of the tensile modulus (about 6.5msi) are about the best that can be achieved in uni-directional laminates (see "Fiber geometry," below).

Partly, this loss of strength occurs because a certain amount of damage from handling is naturally inevitable. Twisting the strands to form yarns also reduces strength. And even the most careful methods of production may leave some fibers "dry"—i.e. not totally surrounded by the plastic resin—and so unable to share the task of load-bearing with other fibers.

Mostly, though, the reduction in properties in a finished part results from the fact that a large part of any composite is the plastic resin matrix, which contributes little to strength. Still, properly fabricated E-glass laminates, while inferior to metals on a stiffness-to-weight basis, compare very favorably in terms of strength-to-weight. The Corvette's GFRP springs are made with E-glass fibers.

S-glass & S-2 glass

Developed during the early sixties for high-performance aerospace and military applications, "S-glass" (a trademark of Owens Corning Fiberglas), containing more silica than E-glass, and no boron, was as much as one-third stronger than its predecessor, having a single filament tensile strength of 685–700ksi. With an elastic modulus (Young's modulus) of about 12.6msi, it was also more than 20 percent stiffer.

Unfortunately, the high cost of certification for aerospace and military applications, and the low rate of production, meant that S-glass cost more than ten times as much per pound as E-glass. To gain the benefits of S-glass at lower cost, a commercial version was developed during the late sixties. The result, "S-2 glass" (a trademark of Owens Corning Fiberglas), has the same chemical composition as S-glass, and essentially the same mechanical properties, but costs much less than the early S-glass, though still 4-5 times as much as E-glass.

S-2 glass fiber is used in the manufacture of floor panels for commercial jet aircraft, for helicopter rotor blades and for filament-wound pressure bottles (see "Fiber geometry," below). One significant automotive application is in the chassis structure of a new commercial vehicle being made by Consulier—builders of the exotic sports car of the same name.

Carbon (graphite) fiber (CF)

During the past thirty years, new fibrous materials have been introduced that match or exceed the strength-to-weight numbers of traditional materials, including GFRP, and vastly exceed their stiffness-to-weight values. Of these, carbon fiber came first.

One method of production starts with a filament of polyacrilonitrile (PAN), similar to ordinary synthetic sewing thread, which is then baked in a high-temperature furnace. During this "pyrolyzing," the PAN is transformed into a 0.0005in diameter filament of nearly pure carbon. A newer, and lower cost method of production starts with pitch (from coal tar), instead of PAN. Depending on the details of the pyrolyzing process, the resulting product can be tailored toward maximum stiffness or maximum strength.

Tensile strengths as high as 600ksi have been claimed for high-strength (HS) carbon fiber, and moduli as high as 100msi claimed for high-modulus (HM) fiber. As with all fiber reinforced plastics, the properties of the finished laminate fall far short of the figures attained for individual fibers. Nevertheless, tensile strengths of 250–255ksi for uni-directional laminates and a modulus of 21msi are cited by one supplier. When it is considered that finished laminates have a density around 0.06lb/cu-in—about the same as magnesium alloys—the possibilities of these materials can be appreciated.

By the mid-seventies, carbon fiber had begun to carve out a significant niche for itself in the aerospace field, including rudder sections for the DC-10, and a virtually all-composite wing on the

The underbody of a Formula One car looks as if it is just an unstressed body panel, but the aerodynamic forces can be very large indeed—a ton or more at high speeds! Though it is expected in chassis "tubs" and airfoils, the use of carbon fiber in the illustrated application is not overkill. *Forbes Aird*

AV8-B Advanced Harrier. Carbon fiber composites were used as early as 1970 in wings for Formula One cars, and the McLaren MP4—introduced at the end of 1980—was the first F1 car with an all-carbon-fiber composite chassis. By 1985, virtually all top ranking F1 teams were making their chassis tubs from composites, principally carbon fiber. CF is also used to make race car wheels, engine push-rods, truck springs, and drive shafts.

Aramid ("Kevlar") fiber

Aramid fiber (marketed as "Kevlar" by DuPont de Nemours, Inc.) was first introduced in 1972. It immediately found application as a "stronger than steel" reinforcing cord for tire construction and, more recently, has been used as a replacement for asbestos in clutch and brake linings. Kevlar, particularly high-modulus "Kevlar 49," is also found in high-performance composites.

Though less stiff, some forms of Kevlar are even stronger than carbon fiber, at least in tension. When pushed on rather than pulled, however, a Kevlar laminate's behavior is unique—it fails by yielding, much like metals. This phenomenon limits a Kevlar laminate's compressive strength to relatively low values, compared to carbon.

Significantly, a substantial amount of energy gets soaked up when a Kevlar laminate collapses in this way, and even after "failing" in compression, it retains a useful fraction of its original strength in tension. This, plus its general toughness, makes Kevlar attractive to manufacturers of lightweight structures such as aircraft and race cars. It accounts for ten percent of the empty weight of DeHavilland's Dash 8 commuter aircraft, for instance. And although the amount of Kevlar in Formula One car chassis is less now than half a dozen years ago, it still sees applica-

For most race car parts made from glass fiber and epoxy, aramid fiber can be substituted for the glass. This makes for a lighter yet somewhat stiffer and stronger part, although at a cost premium. *Forbes Aird*

tion (precisely because of the way it fails in compression) as a hedge against the brittleness of carbon—a matter of considerable concern, both in crashes and at bolted connections.

Apart from its use in high-performance composite structures, Kevlar is widely used for lightly stressed body panels on race cars and similar applications, simply because of its low density. For such applications, Kevlar cloth directly substituted for glass fiber cloth of the same thickness yields a part that is at once stronger and lighter.

Kevlar 29

Individual strands of "Kevlar 29," impregnated with epoxy resin, have a tensile strength of about 550ksi, and a tensile modulus of about 12msi—values that compare closely with S-2 glass. Note, however, that Kevlar (at 0.052lb/cu-in) is about 40 percent less dense than S-2 glass (0.090lb/cu-in). The tensile strength and stiffness is thus at least 50 percent greater than S-2 glass, per unit weight. Rope and cordage, and lightweight body armor are major applications Kevlar 29.

Kevlar 129

A recent variation on Kevlar 29 has been introduced which offers an increase of about 15 percent in strength, and an improvement in stiffness of over 30 percent. Since Kevlar 129 fails in compression the same way as other aramids, its greater strength means it absorbs more energy when it buckles. This has led to lighter, tougher body armor, and recommends it for consideration in expendable structures for race car crash safety.

Kevlar 49

High-modulus "Kevlar 49" (impregnated strand tensile strength of 420–525ksi; modulus of 16.3msi) finds its way into advanced composites in aircraft, aerospace, and race cars, as well as lightweight canoes and kayaks. Properly made uni-directional composites of Kevlar 49 tested in tension are four to five times as strong as steel, on a per-weight basis. While inferior to HS carbon fiber laminates in that regard, they are generally superior to laminates made from HM carbon fiber.

Composites made from Kevlar 49 compare less favorably to carbon fiber composites in terms of stiffness, both on a per-volume and a per-weight basis. Stiffness-to-weight is about one-half and one-quarter the values for HS and HM carbon fiber laminates, respectively. In compression,

too, Kevlar 49 falls far short of the numbers associated with CF. Nevertheless, used in a uni-directional laminate, it equals 7075 aluminum alloy in weight-specific compressive strength, and outperforms 2024 aluminum by two-to-one.

Kevlar 149

In the process of manufacturing aramid fiber, the properties can be adapted toward maximum stiffness or maximum strength. One result of such tailoring is Kevlar 149, which provides an increase in stiffness of up to one-quarter compared to Kevlar 49 (20.7msi vs. 16.3msi), in exchange for a corresponding reduction in strength (about 330ksi vs, say, 435ksi). While Kevlar 149 is still made, DuPont admits to its being "de-emphasized," so it will doubtless prove hard to find.

Surface Treatment & Coupling Agents

To prevent damage to the fibers, whatever the raw material, and to reduce abrasive wear on weaving looms and other processing equipment, it is usual to coat their surface with a "binder"—often a mixture of oil and starch. Since such a lubricant would prevent proper adhesion between the fiber surface and the resin matrix, the binder must be removed before the fabric is incorporated into a laminate. In the case of glass fibers, the binder is burned off, yielding a "heat cleaned" surface.

A freshly made laminate using heat cleaned glass fabric, in combination with any of the common low-temperature resin systems (see "Resins," below), would appear satisfactory at first. Successive tests over time, however, would reveal a gradual loss of strength, as the bond between the fiber and the resin deteriorates. To ensure a permanent strong bond, a "coupling agent" is applied to the heat cleaned surface. The "finish," as these coupling agents are termed, is chosen according to the type of resin to be used. Common glass fiber finishes suitable for use with polyester, vinylester and epoxy resins include "Volan A" and "Garan."

Heat cleaning cannot be used for carbon or aramid, as the extreme heat would damage or destroy the fibers. In the case of aramid cloth, the binder is cleaned by washing, producing a "scoured" fabric. In the case of carbon, the fibers are given a very thin surface coating of epoxy resin, which stays on throughout all subsequent operations.

DIRECTIONAL PROPERTIES OF UNI−DIRECTIONAL FABRIC LAMINATES

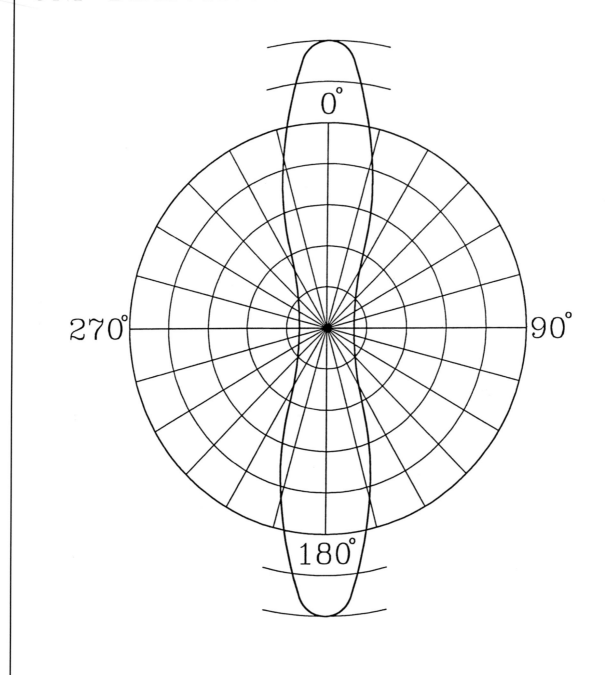

In this diagram, the strength in various directions is represented by the distance from the center of the circles. At plus-or-minus 90 degrees to the yarn axis, what is being measured is almost entirely the strength of the resin matrix.

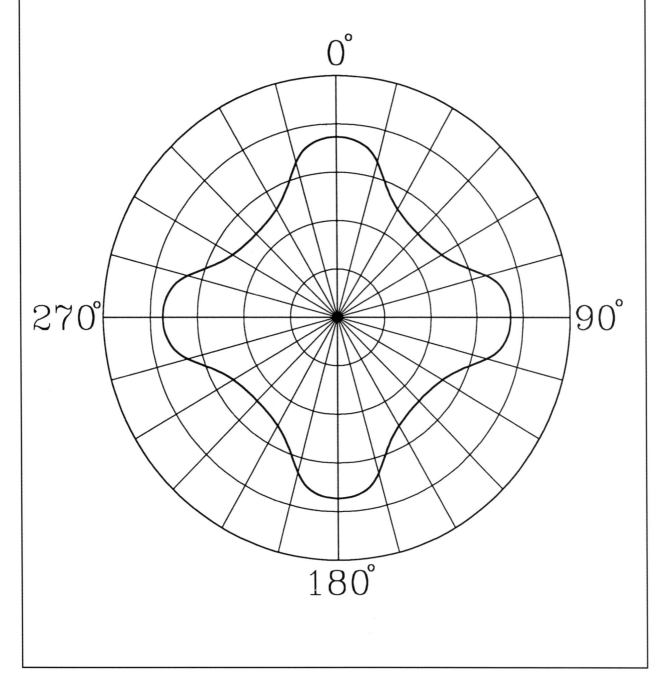

DIRECTIONAL PROPERTIES OF 'BALANCED' FABRIC LAMINATES

A fabric weave with the same number and weight of yarns running in the warp (lengthwise) and fill (crosswise) directions will produce a laminate with the same strength and stiffness in both directions. Note, though, that the values are not equal in all directions—at angles between 0 and 90 degrees, properties are reduced somewhat.

Fiber Geometry

Fibers are one thing; a structurally useful piece of material is quite another. To produce a rudder section for a DC10, or a chassis "tub" for a McLaren, or a wing for a sprint car, you need sheets of the stuff.

Woven Fabrics

One approach is to gather or twist the fibers into strands, then weave multiple strands into a fabric. Glass fiber cloth will be familiar to most readers; Kevlar and carbon fabrics look much the same, except for their color (aramid is yellow-or-ange, carbon fiber is black), and may be handled similarly.

A dizzying array of fabrics is produced by numerous independent weavers. The fabrics vary widely in: the number of filaments in each yarn; the twist of the yarn; the tightness of the weave; and whether the yarn count is the same for both warp and fill (giving a "balanced" or "bi-direction-al" fabric), or consists almost entirely of yarns running in one direction, producing a "uni-directional" fabric, etc. For any given type of fiber, though, the factors of greatest significance are the volume fraction of fiber and the type of weave.

Open weaves make it easy to wet out the cloth, but result in a laminate with lots of weak, heavy plastic. Tight weaves are more difficult to saturate, but give the largest volume fraction of reinforcement and so the best strength and stiffness values. "Balanced" weaves give similar properties when measured along the warp direction and at 90 degrees. (Note, though, that the behavior is not truly "isotropic"—the same in all directions; the strength and stiffness is lower when measured at angles between 0 and 90 degrees than when the load is aligned with the yarn.)

"Uni-directional" weaves, as you would expect, show better performance than balanced weaves when measured along the principle axis, but are very weak in the opposite (90 degree) direction.

Satin and Other Weaves

The crimping of the yarns that occurs as they cross over and under each other tends to reduce the tensile and, particularly, the compressive strength of woven fabrics, compared with a straight yarn. Also, it is difficult to get balanced fabrics to drape properly into inside curves, especially ones with complex changes in radius.

"Satin" weaves, in which each warp thread crosses over several fill threads before crossing under one, reduce the crimping of the yarns and so generally give slightly superior strengths, especially in compression. Satin weave fabrics also have excellent "drapeability," but are somewhat more expensive than balanced fabrics.

Woven Fabric Limitations

Cloth woven from high-performance fibers is widely used in both aerospace and automotive applications, but woven products have some formidable drawbacks. First, the unavoidable spaces between warp (the yarns that run lengthwise) and fill (the crosswise yarns) result in a finished product that is short on strong fiber and long on weak resin. About 60 percent fiber volume fraction is the limit for woven fabric laminates.

Second, as already noted, each strand within the fabric is "pre-buckled" as it snakes through the weave, markedly reducing the compression strength of the finished laminate. Despite these liabilities, woven products are used for parts requiring strength in many directions and for shapes too complex for the alternative—uni-directional fabrics.

Uni-directional Fabrics & Tapes

The very best mechanical properties are achieved with uni-directional fabrics, in which all the fibers run in the same direction and are packed closely side-by-side, although a few, light "fill" yarns may be woven at intervals along the "warp," for convenience of handling. The close packing, the freedom from the kinks inherent in woven goods and, above all, the uni-directional alignment of the fibers, yields the strongest and stiffest laminates. With the best fabrication techniques, over 75 percent fiber volume fraction can be achieved with uni-directional fabrics.

Obviously, such uni-directional materials have strength and stiffness in only one direction—the material has a grain, like wood. Useful bi-directional performance is achieved by the "plywood" method—alternate layers applied at an angle to one another.

Filament Winding

Three centuries ago, barrels for top quality guns were manufactured by a process known as "Damascus twist"—successive layers of iron wire were wound onto a mandrel, and "hammer weld-ed." A similar process is used to produce fiber reinforced plastic goods, such as storage tanks, golf club shafts, ski poles, vaulting poles, and miles of pipe for all purposes. In the automotive field, wheel rims, engine push-rods, drive shafts and other cylindrical parts are made in this way.

While straightforward mechanical means, such as a lead screw, can be used to guide the fibers onto the mandrel surface, such simple mechanisms limit the possible patterns of fiber application. For precise control over the angle at which the fiber is deposited, sophisticated electronic controls are needed for the independent drives of both the rotating mandrel and the guide head that lays down the filament. Some filament wound products contain up to 90 percent fiber.

Careful control also needs to be maintained over the tension in the filament as it is applied to the mandrel. Depending on the geometry of the part, the winding pattern, and whether the winding is done "dry" or "wet," there may be a tendency for the filament to slip out of position on the surface. At the same time, tests show that residual stresses in the fibers after curing, resulting from the tension maintained during winding, has a marked effect on the strength of the finished part.

Filament winding is not restricted to perfectly cylindrical shapes. Any "surface of revolution"

For each material, though, there is some critical crack depth that can be tolerated, so if the material is drawn out into filaments thinner than this critical dimension, the mere fact that a fiber is continuous pretty well guarantees that it is free of surface cracks.

---

can be produced by winding, although sharp outside corners create problems, and undercuts must be avoided. In addition to circular shapes like rocket motor cases and torpedoes, helicopter rotor blades are one example of non-cylindrical filament wound profiles produced by the aerospace industry. To the writer's knowledge, no one has yet produced a complete chassis "tub" in this way, but the concept is intriguing because of the opportunity to optimally orient the fibers in relation to the applied loads.

Pultrusions

As the name suggests, pultrusion is a process like extrusion, except that the material is pulled, rather than pushed. Bundles of resin-saturated fiber are pulled through a heated die, which both shapes the finished product and effects the cure. Virtually any cross-section can be produced, including hollow sections, but obviously only straight goods can be made in this way.

Like filament winding, pultrusion uses the lowest cost raw materials (unwoven continuous filaments or strands), and is well suited to continuous mass production. Depending on the shape of the part, pultrusion and filament winding can be combined to control strength properties lengthwise and across the section.

Apart from industrial goods like ladder rails, electrical insulators and reinforcing rods for concrete, pultrusions are used in automobiles for anti-intrusion door beams, bumper beams, etc.

Mat

Parts with very complex shapes and/or loads may be impractical to produce using any continuous fiber reinforcement. In such cases, a felt-like mat may be chosen, consisting of fibers chopped into short lengths, held together with a resin-soluble binder.

Unlike woven fabrics, laminates made from mat exhibit equal, but low, strength and stiffness in all directions. The low strength and stiffness is partly because of the short fiber length, but mostly because the same random orientation of the fibers that provides the isotropic mechanical properties also prevents close packing of the fibers. As a result, mat laminates seldom contain more than about 30 percent fiber.

Chapter 9

Plastic Composites—2: Matrix Resins

The discussion of high-strength fibers in the previous chapter might suggest that, for useful component strength, the individual fibers have to extend from one end of a composite part to the other. But remember that a composite has two components—fiber plus matrix.

Reinforcing fibers can be embedded in virtually any matrix material that can exist in both a fluid and a solid state, including ceramics, metals, and plastics. Ceramics and metal matrix composites (MMC) are discussed in chapter 12; our interest here is plastic matrices.

The low stiffness of the plastic resins used as matrices makes them next to useless as structural material by themselves yet, paradoxically, it is that very "rubbery" nature that allows each fiber to "hand off" the load to adjacent fibers. Without a matrix surrounding them, high-strength fibers would be useful only for ropes and similar tension members—it is the matrix that allows us to take advantage of the properties of strong, stiff fibers in structurally useful composites.

In a sense, the reinforcement provides the strength; the matrix provides the shape.

'Theory of Combined Action'

In Part 1, we saw that a sample of material which is pulled lengthwise will stretch somewhat. The amount of stretching (the strain) depends on the load per unit of area (the stress) and on the elastic modulus of the material. And this works both ways—we can calculate the stress on the basis of the strain.

For example, if a steel bar of 1sq-in cross-section extends under load by 1 percent, we can easily calculate that the stress in the bar is 300ksi (300,000psi). Similarly, if a bar of identical size, but made of rubber, extends by 1 percent, the same calculation would reveal that it was stressed at just 10psi! Now imagine the steel rod surrounded by a uniform layer of rubber 0.234inch thick, so that the cross-sectional area of the rubber is also 1sq-in. If that composite of steel and rubber extends by 1 percent, the internal stresses must be in the ratio of 30,000 to 1.

Similarly, the stiffness of the plastic resins used in composites, while much greater than that of rubber, is still profoundly less than that of the fibers. Clearly, the matrix of a composite is lightly stressed, even when the fibers are carrying a very large load, and since the area of contact between the fiber and the plastic extends over the whole surface of the fiber, the matrix has no problem transmitting the load from one fiber to another.

The matrix contributes some other vital properties to a composite. As just suggested, no matter how strong and stiff something shaped like a rope may be when you pull on it, it buckles when you try to push on it. Another function of the matrix, then, is to stabilize the fibers against buckling—in a sense, the reinforcement provides the strength; the matrix provides the shape. The matrix also screens the fibers from harmful factors in the environment, and maintains their integrity by protecting against scratches.

But if the matrix protects the fibers from surface damage that could cause cracks, what protects the matrix against cracking? Why, the fibers, of course! Any surface crack in the resin that tries to advance deeper into the material soon runs into a fiber; to progress, the crack needs to "go around" the fiber. But the fiber is

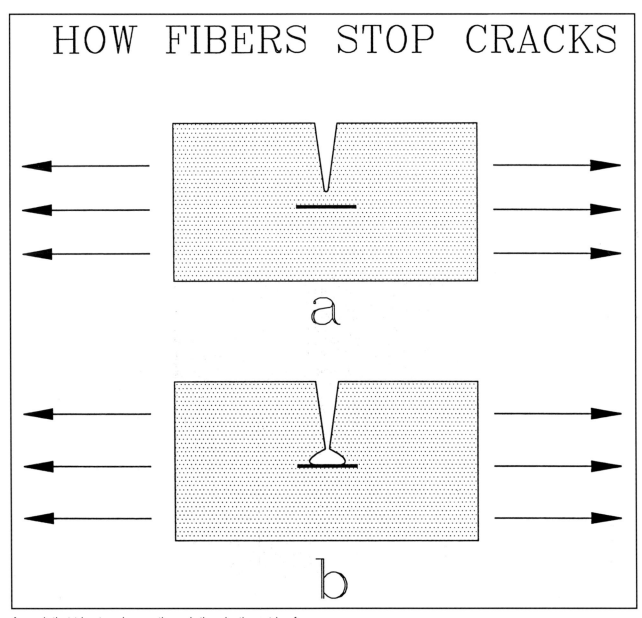

HOW FIBERS STOP CRACKS

A crack that tries to advance through the plastic matrix of a composite soon runs into a fiber. For the crack to progress past the fiber, it has to grow wider at the leading edge, effectively "blunting" itself.

very long, compared to the radius at the tip of the crack. To advance, then, the tip radius has to grow larger, effectively blunting the crack, in somewhat the same way as a hole drilled at the end of a crack will serve as a "stop."

Among the plastics, we can distinguish between thermoplastics (see chapter 7), which soften when heated and harden when cooled, and thermosets, which change irreversibly from liquid to solid as a result of a chemical reaction. Fiber reinforcement, usually glass, is often added to thermoplastics like nylon to increase their strength and stiffness. Even so, the structural performance of such parts is not sensational, and making them is a

But if the matrix protects the fibers from surface damage that could cause cracks, what protects the matrix against cracking? Why, the fibers, of course!

CARBON FIBER Wheel

Fiber-reinforced thermoplastics are uncommon in high-performance applications; those using carbon fiber for the reinforcement are even more rare. MRT offers a variety of race car wheels—mostly for sprint car racing—in thermoplastic (described as "nylon-like"), reinforced with CF. *MRT*

complex industrial operation, well suited to mass-production but impractical on a one-off basis. Truly high-performance composites invariably use a thermoset plastic matrix, almost always of polyester, epoxy, or vinylester.

Resins
Polyester

By far the majority of low-pressure laminating, including boats, furniture, autobodies, truck cabs, etc., uses polyester resin. Polyesters are inexpensive and easy to work with, but result in laminates with mechanical properties inferior to those made with the more costly epoxy or vinylester resins. Part of the reason for this is polyester's shrinkage as it cures—un-reinforced polyester will reduce in volume as much as 3 percent in changing from liquid to solid. Note that this is 3 percent by volume; the linear shrinkage will usually be less than 1 percent.

That does not mean, however, that an 8ft long race car body part will wind up an inch shorter than the mold it was made in! In a laminate containing a significant amount of reinforcing fiber, the compression stiffness of the fiber will resist the polyester's attempts to "get small." That fight between the matrix and the reinforcement, however, leaves the resin pre-

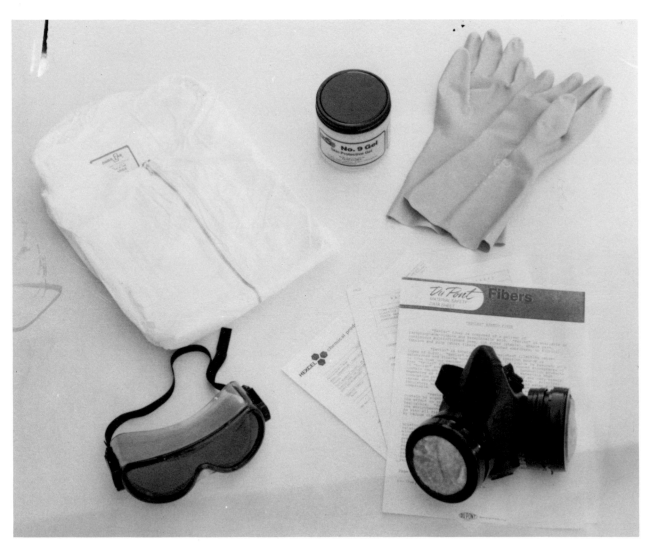

The resins used to produce fiber-reinforced-plastic composites and their fumes, as well as fiber fragments and dust, can present significant health hazards. Virtually complete protection is afforded by adequate ventilation, when combined with five essential items: respirator, gloves, barrier cream, disposable coveralls, plus the product manufacturers' Material Safety Data Sheets. *Forbes Aird*

stressed in tension, which reduces its apparent tensile strength. Polyester also does not adhere well to aramid fiber, and so should not be used with "Kevlar."

A couple of different polyester resin types are sold. Both "ortho" (orthopthalic) and the slightly more expensive "iso" (isopthalic) resins are produced in huge quantities by numerous suppliers. They are syrupy liquids varying in color from milky white to straw.

Polyester resin, as usually purchased, actually contains an assortment of other chemicals. Styrene monomer—also known as vinyl benzene—accounts for about 40 percent of the mix, and for most of the familiar fiber glass shop smell. It is a clear, colorless liquid which serves as a diluting solvent to reduce the viscosity of the raw polyester (as an aid to wetting-out the reinforcement), and also reacts with the polyester during the curing process.

CAUTION: Styrene is highly flammable, and its vapors are explosive. Contact with the liquid can cause dermatitis. The vapors are irritants to the eyes and respiratory tract. Inhalation can cause headache, dizziness, and vomiting; high concentrations cause central nervous system depression.

Dimethyl aniline (DMA) and/or cobalt naphthenate (CoNap) are usually present in store-bought polyester resin as accelerators, or "promoters," to speed up the curing process. DMA is supplied as a clear, light-yellow to brownish solution.

CAUTION: DMA is strongly alkaline and so can cause skin burns. Both liquid and vapor are a central nervous system depressant, and inhalation of the vapors may be fatal. DMA vapors are flammable.

Cobalt naphthenate solution is a red-violet liquid, with a mineral spirits odor. (The slight purplish cast often seen in polyester resins implies the presence of CoNap.)

CAUTION: Cobalt naphthenate solution is combustible. As liquid or vapor, it is irritating to the eyes and skin. Breathing of concentrated vapors may cause headache and loss of coordination. It is especially harmful if swallowed.

To convert the liquid resin into a hard solid, a catalyst, usually methyl ethyl ketone peroxide (MEKP) or benzoyl peroxide (BPO), is added by the user at the time of processing. The rate of curing can vary over a wide range, according to the size of the part, the temperature, and the resin/catalyst ratio.

MEKP is a clear, colorless liquid, and is explosively unstable stuff in its pure form. As a result, it is supplied dissolved in a buffering liquid. BPO is seldom used for curing polyester and vinylester based laminates, but is the common catalyst for polyester-based body fillers. It is usually supplied as a paste.

CAUTION: While MEKP is not highly flammable, it is a powerful oxidizer which may cause paper, rags, etc. to spontaneously ignite. It is also corrosive, and contact may cause skin burns. MEKP splashed in the eyes can cause serious, permanent eye damage. The vapors may cause headaches and intoxication, and corrosive damage to nose, throat, and lungs.

In any commercial polyester resin, there may also be a wide assortment of fillers, inhibitors, pigments, and plasticizers, to obtain special properties. "Gel coat" is an example of a specialized formulation of polyester resin.

Gel coat

For parts such as automotive bodywork, where a smooth, shiny surface is required, a layer of resin is sprayed onto the mold surface and permitted to cure before any reinforcement is applied. This "gelcoat" is simply a polyester resin with an inert filler added to give the resin enough body (or "thixotropy") to resist running and sagging, and is pigmented to be self-colored. It provides a hard, shiny, attractive surface finish on the exposed side of the part, which resists weathering and requires no maintenance.

Epoxy

Variations in the strength of the resin have a relatively small effect on the mechanical properties of the finished part. Nevertheless, an improvement in strength and stiffness results from the use of epoxy in place of polyester, and some fibers and finishes are incompatible with polyester. On the other hand, epoxies are rather trickier to handle than polyester, and also have different health hazards associated with them. They are also significantly more expensive—in drum quantities about twice that of polyester, and when allowance is made for the more expensive curing agents, the total can rise even further to over $2.50 per pound.

Epoxy resins used in composite fabrication are moderately viscous liquids, varying in color

Truly high-performance composites invariably use a thermoset plastic matrix, almost always of polyester, epoxy, or vinylester.

from near water-clear to dark amber. Unlike polyester and vinylester, the cure rate of epoxy systems cannot be adjusted by varying the catalyst/resin ratio. For any given combination of resin and hardener, there is a narrow range of proportions which will yield a proper cure. Adjustment of the cure rate requires altering either the temperature or the curing agent used.

Liquid epoxies, and appropriate curing agents, are available which cure at room temperature. The highest performance laminates, though, are usually produced using epoxy resin/hardener systems which are formulated for elevated temperature cures, and this requires an oven of some sort. Even room-temperature curing epoxy systems benefit—especially in terms of their mechanical properties at higher temperatures—from a post-cure heating. (One supplier of "kit-plane" plans recommends painting the structure black and leaving it out in the sun!)

CAUTION: Skin contact with either the liquid or vapors of some epoxy formulations can cause moderate to severe irritation, and inhalation of their vapors can cause irritation of the respiratory tract. Ingestion of even the smallest amounts can lead to an overall allergic response.

Among the more common hardeners for room-temperature cures are modified amines and polyamides. Amine curing agents are low in viscosity, and are highly active—a relatively small amount (typically less than 25 percent of resin volume) does the job. Conversely, they are intolerant of variations in the resin-to-hardener ratio.

CAUTION: Industrial formulations of amine curing agents (not generally available to the public) can cause dermatitis, severe burns, and may have other serious long-term health effects. Di-

rect contact with the eyes will produce severe damage. While the modified amine hardeners available to the public have very much reduced potential for both irritation and sensitization, prolonged contact with either liquid or fumes may cause both effects in some individuals.

Polyamides are viscous liquids with a distinct odor of ammonia. They will tolerate large variations in the ratio of resin-to-hardener, and are probably safer than any of the amine curatives. On the other hand, they are uselessly thick unless diluted. Dilution to workable viscosities is achievable, but excessive diluent may result in a structurally inferior end-product.

Vinylester

About twenty years ago, a new family of resins was introduced that resemble polyester in their ease of handling, yet yield mechanical properties near to those of epoxy. (The vinylester molecule is usually described as being like epoxy at one end and like polyester at the other.) In price, too, these vinylester resins fall between polyester and epoxy.

Vinylester resin systems, like polyester (and for the same reasons), include large amounts of styrene in the can as-bought. They also generally use the same catalysts and accelerators as polyesters. When it first became available, vinylester was supplied without an accelerator pre-mixed into the resin. This limitation has now been dealt with, and vinylesters can be handled very much the same as polyester.

Pre-pregs

Whatever combination of resin system and reinforcement may be selected, the traditional "bucket-and-brush" technique of saturating the reinforcement limits the choice of resins, and the hand work introduces unpredictable variations in properties.

The preferred technology (now used widely in F1 tubs and almost exclusively in the aerospace field) employs "pre-pregs," in which the fibers are uniformly machine-impregnated with resin and then partially cured, the process being arrested by refrigeration. The fabricator need only trim these leathery sheets to size and shape, drape them into (or onto) the mold, and complete the cure by applying heat. Apart from sparing the fabricator the mess of liquid resins, pre-pregs offer superior resin systems, and much more consistent results.

The resin systems employed are epoxies, but of a type which can be interrupted at what is termed a "B-stage" cure. Elevated temperature cures are used, because room-temperature curing resins could not be arrested for a usefully long time, and because the high-temperature systems yield the best mechanical properties.

A wide variety of patterns of woven fabrics, unidirectional fabrics and tapes, plus pre-impregnated yarn and roving are available from a number of suppliers, in E-glass, S2-glass, Kevlar, HS carbon, and HM carbon, as well as hybrid fabrics using combinations of these reinforcements.

Limitations of Composites

For all their benefits, composites are not without problems. When structural metals are overloaded, their ability to yield provides a degree of insurance against sudden, total failure. With the exception of aramid fiber composites loaded in compression, composites (and other non-yielding material, like wood and concrete) lack this "user-friendly" characteristic.

Second, the directional nature of composites presents a cleft stick to the designer. It permits aligning the material with the applied loads, for maximum performance, but requires that the designer know the exact direction the loads come from! It also means that when loads come from many directions, the need to provide fibers running in multiple directions drops the apparent

---

The highest performance laminates, are usually produced using epoxy resin/hardener systems which are formulated for elevated temperature cures, and this requires an oven of some sort.

---

specific strength and specific stiffness of composites from incredible to merely sensational.

Cost is a factor, too. Not only are the materials themselves expensive, there are also the costs of labor and of capital invested in equipment to think about.

Traditional "fiber glass" work requires negligible "plant" and, by substituting higher performance fibers, laminates that are structurally competitive with metals can be achieved with equally simple equipment, justifying a considerable premium for the material. Besides, very little of it is used. Nevertheless, for the really big strength and stiffness numbers, some pretty specialized gear is needed, including refrigerators for storing pre-pregs, and large autoclaves (think pressure cooker) for heat-curing.

Joining composites can also prove problematic. While "co-cured" assemblies can avoid many jointing problems, there are many applications in which it is necessary to produce numerous separate parts and then somehow connect them together. If "point fasteners" (bolts, rivets, or screws) are used, the local stress concentrations require additional material to be added—lots more in the case of carbon fiber composites. (This was probably a major factor in the commercial failure of the Learfan aircraft—the added weight of the extra material at joints, plus the weight of the fasteners themselves, resulted in a structure that could have been duplicated in metal for about the same weight.) And if point loads are avoided by the use of adhesive bonding, those "secondary bonds"—the ones achieved after the joined parts have cured—require special pre-treatment, and their security cannot be confirmed by ordinary visual inspection.

There is also the problem of quality control—the composite manufacturer is making not just a structure but also the material from which it is fabricated. In aviation and aerospace work, a high level of consistency is ensured by constant testing of both incoming materials and finished parts, or test "coupons" cut from them. It is not prohibitively expensive to have such tests performed, and fabricators of one-off or limited production parts should consider that, without these tests, any assumed material properties are essentially guesswork.

Chapter 10

Sandwich Construction

A few race car parts are loaded in pure tension; the usual examples are things like rod bolts and head studs. Most places, though, the working loads are more complex, and involve compression or shear stresses at least some of the time. A Panhard rod, for instance, is in tension when turning in one direction, but in compression when turning the other way; shear, compression, and bending stresses predominate in parts like wheels, bell-housings, and, most significantly, in the "stressed-skin" monocoque tubs used in Indy Car and roadrace cars.

In chapter 2 we described how slender structural elements—long tubes, sheets, and thin plates—tend to buckle when loaded in compression or shear, and that it is the ability to resist buckling, rather than strict strength requirements, that determines the minimum weight of the part. We also suggested that the extra weight that has to be carried to meet this need for stiffness, as compared with strength requirements, is greatest when the loads are comparatively small in relation to the length over which they have to be carried. About the most extreme examples of such stiffness-critical parts in race cars are airfoils (wings) and monocoque chassis. Skins thick enough to have the stiffness required for buckling resistance would be vastly over-strong... and overweight.

What's obviously needed is some way to increase a panel's buckling resistance without proportionally increasing weight. Aircraft have demonstrated one solution—a grid of stiffening frames that reduce the span over which the compression or shear forces have to be carried by dividing one large panel into a number of smaller ones. The ribs and stringers inside wings and the bulkheads in monocoque tubs are typical examples. These frames have some weight, of course, so a choice exists between use of a comparatively thick skin with a few frames, or a thinner skin supported by many closely spaced stiffeners. Carried to its logical conclusion, the second approach leads to *continuous* stiffening, where the stiffeners extend over the whole area of a panel made from skins just thick enough to carry the "real" loads without failing in "true" compression or shear.

The practical realization of continuous stiffening is "sandwich" construction, in which two thin skins are spaced apart by a low-density

The core acts as uniformly distributed framing that stabilizes each skin against buckling, forcing it to face up to its job of load bearing.

Facings for Sandwich Construction

In sandwich construction, the core frees the skins (or "facings," as they are often called) from lateral bending loads that would cause them to buckle. Since the skins can therefore be worked to a much higher fraction of their potential compressive strength, the materials used in the skins of a structural sandwich generally need to be strong and stiff.

Part of the task of the sandwich designer, then, is selecting among available core materials on the basis of density, so as to strike the optimum trade-off between the relative weight and thickness of the core and that of the facings.

Traditional metal airframe structures use ribs and stringers to divide large areas into smaller ones, reducing the tendency to buckle under compression of shear loads. *Forbes Aird*

filler. The core acts as uniformly distributed framing that stabilizes each skin against buckling, forcing it to face up to its job of load bearing. This can be a very effective way to increase panel stiffness at light weight. Note that a sandwich panel is not a material—it is more like a structure itself, and can be made from many combinations of skin and core materials.

Aluminum

Metals are one obvious choice, and metal sandwich chassis construction has a long history in race cars. As early as the thirties, Alec Issigonnis' lightweight Special was built from "Plymax"—thin sheets of aluminum glued to both sides of aircraft grade plywood.

McLaren built a Formula 1 car chassis using a combination of aluminum skins with a core of balsa wood in 1965. (This material has been labeled as "Mallite" in many published works, but the writer can find no record of this trade name. It seems possible that "Mallite" is a corruption or misprint of "Metalite"—a trademark registered by Chance Vought and used in some of their fighter aircraft in the late forties.) Aluminum skins were subsequently used with a variety of core materials in many Formula 1 and Indy cars. Depending on the amount of shaping that has to be performed on the skins, the strength level required, and on availability in the thickness required, many common aluminum alloys may be used, including 3003, 5052, 6061, or 2024 (see chapter 5).

Composites

More recently, sandwich construction using facings made from composite materials (see chapters 8 and 9— "Composites") has become universal in both Indy Car and roadrace car chassis. Partly, this is because of composites' high specific strength and stiffness, compared to traditional metals. (We have already suggested that a well-designed sandwich structure can work the facings quite hard.) Another reason is the comparative ease of forming composites into complex shapes. Also, depending on the intensity of the loads, a metal skin sufficiently thin for the job might turn out to be more like foil—almost impossible to handle without damage. Finally, one of the most convincing arguments for composite skins for sandwiches is the relative ease of joining them to the core.

For the facings to be effectively stabilized, they must be well stuck to the core. Adhesive bonded aluminum, using epoxy as the glue, is commonly employed in both sandwich and conventional single-skin aircraft structures, but ensuring a sound joint demands special pre-treatment of the aluminum surface. Since the composites employed as facings in most sandwich construction (and almost exclusively in race cars) already use epoxy as a matrix material, obtaining a reliable connection between the facings and the core is much easier.

There are two general approaches. The skins can be produced in advance, then adhered to a separately formed core, or the sandwich can be built up in the mold, then the entire assembly "co-cured"—the facings and the adhesive harden at the same time. (Alternatively, one facing and the core can be co-cured, then the second facing applied.) The choice among these methods depends on the details of the matrix resins used, whether the cure cycle involves heat or pressure which might damage the core, whether both inner and outer facings need a perfectly accurate molded finish, and other factors, including plain old convenience.

Wood

We suggested in chapter 2 that there is a general relationship between the overall size of a structure and the magnitude of the loads it has to bear, on the one hand, and the density of the material from which it is made, on the other hand. Accordingly, some lightly loaded structures can profitably use low density facings. A classic example of this is the World War II deHavilland Mosquito bomber, which employed a plywood/balsa/plywood sandwich for making both fuselage and wings. The success of this aircraft confirms the wisdom of its designers. (See also chapter 7)

Other Metals

On the other hand, structures with larger loads in relation to their size sometimes use skins of a more dense material. Ordinary carbon steel is impractical as a facing material, if only because corrosion would rapidly trash the foil-thin skins. But titanium or stainless steel skins, oven brazed or spot welded to a compatible metal "honeycomb" core (see "Cores," below), are used in some aircraft applications, particularly for jet engine tail pipes. Needless to say, this form of construction is hellishly expensive!

Cores for Sandwich Construction

We have already suggested that the core in a structural sandwich stabilizes the skins against

Density of Some Sandwich Core Materials

Material	Density lb/cu ft	Shear strength, psi, @ 6 PCF density
Balsa	6–15	140–170
Polyurethane Foam	4–10	100
Methacrylamide Foam	2–12	250
Paper honeycomb	1.5–8	150–250
Aluminum Honeycomb	1–50	300–500

buckling by carrying the lateral forces that threaten to start the buckling. Loads at right angles to the plane of the facings, of course, act toward or away from the core, so the core has to sustain compression and tension forces. In addition, shear loads arise in the core when the entire panel is subject to either shear or bending. (Many designers operate on the assumption that the core carries *all* the shear stress when the panel experiences bending.) The core also has to help feed the loads into and out of the facings, generally via bolts or rivets, producing high localized stresses at these connection points. All of this points toward a strong, dense core.

At the same time, because the core is usually many times thicker than the facings, it can account for a significant fraction of the total panel weight, which argues for a low-density material. Part of the task of the sandwich designer, then, is selecting among available core materials on the basis of density, so as to strike the optimum trade-off between the relative weight and thickness of the core and that of the facings.

Other considerations in the selection of the core material include heat resistance, cost, ease of working, durability, and compatibility with the adhesive used to attach it to the facings. Styrene foam, for example, is easily shaped using the "hot wire" technique, but dissolves in the styrene monomer used in polyester and vinylester resins (see chapter 9— "Resins").

Urethane foams are strongly resistant to solvents and so can be used with polyester and vinylester resins. On the other hand, they must not be hot wired.

Balsa

Soon after World War II, the recreational boat-building industry switched wholesale to glass fiber reinforced plastic (GFRP), which had many advantages over traditional construction in wood. Stiffness wasn't one of them, though, because of the relatively low specific stiffness of glass fibers. That situation led rapidly to selective application of sandwich construction, using GFRP facings over a balsa core, a trend doubtless encouraged by the success of wartime applications of this combination of materials.

In response to that demand, balsa became

readily available in the form of end-grain blocks (the grain runs the "short" direction), anti-fungus treated, kiln dried, and carefully graded into various densities from 5-16lb/cu-ft. To ease "tiling" the square blocks into or onto curved surfaces, they are also available in the form of blankets, with individual blocks spaced slightly apart and glued to a light open weave "scrim" fabric resembling cheesecloth.

Compared to other potential core materials at the same density, balsa exhibits excellent compressive strength and stiffness (with the grain), but its shear strength is inferior. It will tolerate higher temperatures than many foams (the limit is not the wood itself, but the water content, which boils-off of at 212 degrees F), and is comparatively economical. It is also unaffected by the styrene monomer in polyester and vinylester resins.

Apart from its continuing widespread use in boat-building, end-grain balsa is used in the sandwich floor panels in many commercial aircraft. Since its use in the 1965 McLaren, balsa has been supplanted as a sandwich core material in race car monocoques by honeycomb (see below), but on the grounds of cost, availability, ease of working, and familiarity, balsa still offers many benefits to the do-it-yourself sandwich builder.

Foams

Virtually any material can be foamed, including metals, plastics, glass, and a host of other substances—a foam is merely something full of holes. All that is required is that the material be capable of existing in both liquid and solid form, and then solidified after bubbles are introduced into the liquid. The liquid/solid transformation may be achieved by alternate melting and freezing—as in foamed metals and thermoplastics—or by the curing of a thermosetting resin. The bubbles themselves may contain CO_2, freon, or another gas, either introduced into the liquid from an external supply, or formed by a chemical reaction within the bulk of the liquid.

The mechanical properties of foams generally resemble those of the parent material in solid form, reduced in proportion to the density. This is more generally true of properties measured in shear than in other modes of loading—which is fortunate from the point of view of stress analysis, since shear properties are the crucial ones when the foam is used as the core in a structural sandwich. For sandwich cores, our interest is in foamed plastics.

One of the most common foamed plastics is polystyrene. Polystyrene foam comes in two forms—the white, "expanded-bead foam" used in picnic coolers, and true cellular foam, of which the most familiar may be blue "Styrofoam" (a registered trademark of Dow Chemical), widely used for building insulation. Bead-foam is unsuitable as a structural core material, but cellular polystyrene foam has been successfully used as a core in many light aircraft sandwiches. It can be accurately shaped using a "hot wire" cutter, though only single plane curves can be produced in this way. Three-dimensional curves can be quickly formed by hand with a "cheese grater" type of tool used for automotive body filler, and a remarkably smooth surface achieved with an ordinary mill file. Polystyrene foam densities range from about 1.6lb/cu-ft to as much as 50lb/cu-ft. "Styrofoam SM" used for common insulating boards is about 2.0lb/cu-ft. Polystyrene foam softens noticeably above about 140-170 degrees F, and dissolves in polyester and vinylester resin.

Rigid urethane foams, widely used as packaging materials, are available in densities from 3-60lb/cu-ft, although the grades used for sandwich cores are taken from the lower half of that range. Depending on the details of their processing, the range of service temperature limits is from 175–250 degrees F. Urethane foams are strongly resistant to solvents and so can be used with polyester and vinylester resins. On the other hand, they must not be hot wired.

CAUTION: Exposing urethane foams to excessive heat, including hot wire cutting, creates cyanide gas, which is extremely toxic and may be fatal if inhaled.

Urethane foams are also available as "foam-in-place" formulations consisting of two liquids which, when mixed together, froth up and expand to twenty to forty times their original volume. Considering the inherent adhesive properties of urethanes, this feature seems ideal for production of sandwich structures with complex and unequal contours on both faces. It is impossible, however, to obtain uniform density throughout the part, or even to control the average density of the foam. In addition, the forces produced during the "rise" of the foam are very large and may badly distort or even burst the part... and any conventional low-pressure mold.

Other foams used for sandwich cores are methacrylamide and polyvinyl chloride (PVC).

Methacrylamide foam has superior heat resistance and is strongly resistant to solvents, and both exhibit slightly better mechanical properties than other core foams, though at higher cost. Some versions of PVC foam can be worked into complex contours by careful use of a heat gun.

An ingenious alternative is *syntactic* foam, formed by introducing hollow spheres of one substance into a matrix of an altogether different material. Phenolic and, more recently, glass microballoons—from 1/8in down to a few thousandths of an inch in diameter—have been used as lightweight fillers, to reduce the overall density of liquid resins like polyester and epoxy.

Because the buckling resistance of a thin panel increases as the panel is made smaller, and since each wall of a honeycomb can be regarded as a very small panel, it becomes easy to understand the strength and stiffness of honeycomb under loads that tend to crush or shear it.

Honeycomb

For the highest performance, albeit at the highest cost, honeycomb cores are used. As the name suggests, honeycomb is an arrangement of cells, usually hexagonal in shape, though they can be triangular or other shapes—the general idea can be understood by looking at the edge of a piece of corrugated cardboard. Because the buckling resistance of a thin panel increases as the panel is made smaller, and since each wall of a honeycomb can be regarded as a very small panel, it becomes easy to understand the strength and stiffness of honeycomb under loads that tend to crush or shear it.

The first structural honeycomb, during World War II, was made from ordinary paper, coated with a synthetic resin to make it waterproof. Paper honeycomb is still widely used in the construction of household doors, in boats, and even in aircraft—though the paper used in high-performance applications is more likely to be made from aramid rather than wood fiber.

Paper honeycomb is produced by printing stripes of adhesive at intervals across flat sheets of the material, then stacking multiple sheets together so the glue lines on alternate layers are staggered. Once the glue has hardened, slices are cut off and drawn apart, yielding something similar to Christmas decorations—expanded honeycomb.

The benefits of honeycomb were soon extended by applying the principle to other materials including metals, particularly aluminum. Honey-

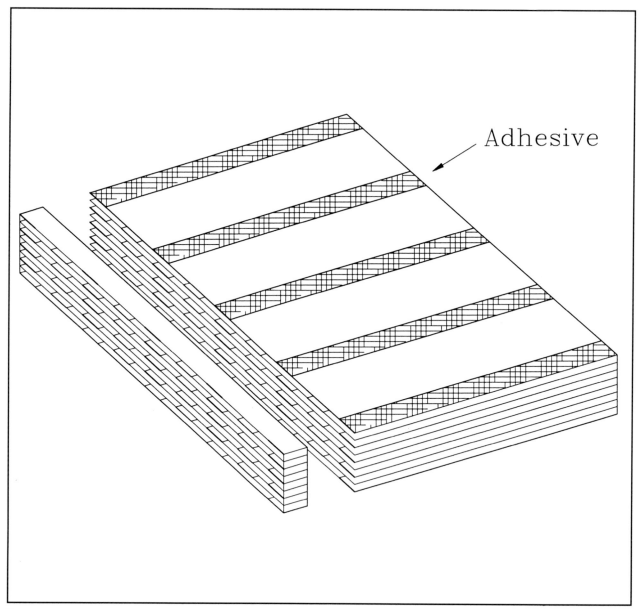

Expanded honeycomb is produced by stacking multiple sheets of material together (often a form of paper made from aramid fiber, but the same technique works with thinner gauges of aluminum), with wide adhesive stripes applied at staggered intervals on alternate layers.

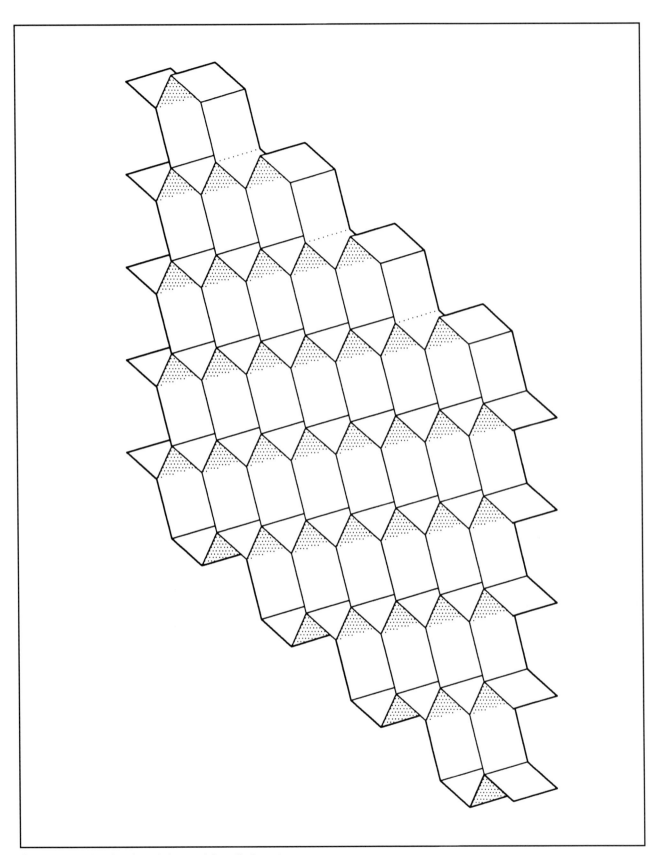

Once the glue has hardened, the stack is pulled apart.

The mechanical properties of foams generally resemble those of the parent material in solid form, reduced in proportion to the density.

comb can be produced from aluminum just the same way as from paper—at least in thin gauges. Aluminum honeycomb is quite stiff so, to facilitate making curved sandwich panels, an "overexpanded" grade of material is produced, with the cells distorted from the usual symmetrical form to a long and thin shape, which allows the honeycomb to drape more readily—although only in one direction.

To produce aluminum honeycomb from sheets of greater thickness, and in higher strength alloys, requires an approach similar to

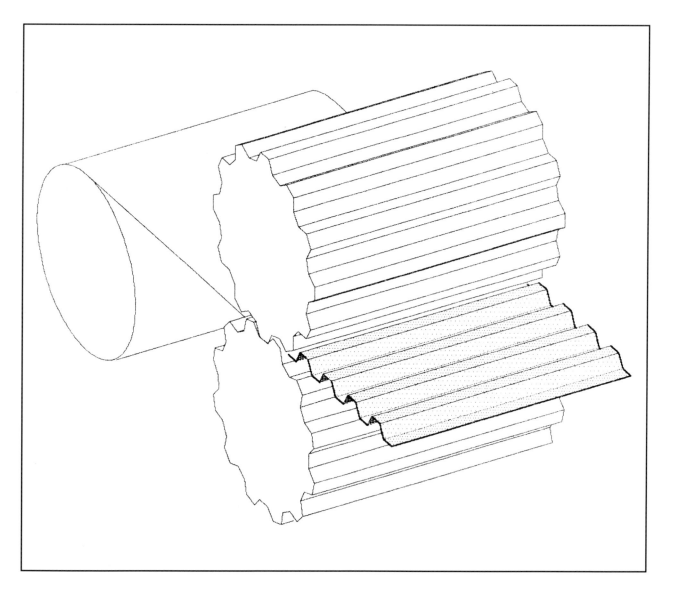

How corrugated honeycomb is made (above). To assist in making curved sandwich panels, and "over-expanded" honeycomb is available, and during manufacture, its cells are distorted (opposite page) to a long and thin shape, allowing the honeycomb to drape more readily in one direction.

making corrugated cardboard—individual sheets are crinkled to produce a "half-cell" shape, the "bumps" covered with adhesive, then successive formed sheets stacked together with a slight off-set. As with expanded honeycomb, the stacks are then sliced, to produce *corrugated honeycomb*.

For composite-faced, honeycomb-cored sand-wiches, the same choice exists between separate forming of skins followed by adhesively joining the complete assembly, and co-curing. Aluminum

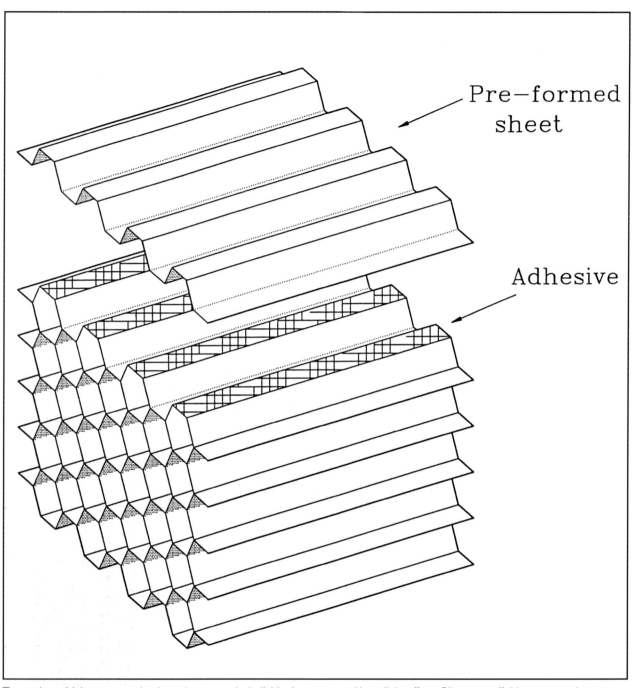

Pre−formed
sheet

Adhesive

To produce thicker gauge aluminum honeycomb, individual sheets are formed into "half-cell" shapes, the high spots covered with adhesive, then multiple sheets stacked togeth- er with a slight offset. Slices cut off this array produce *corrugated honeycomb*.

skin/aluminum honeycomb core sandwiches are usually produced using a thin film adhesive.

Results of Sandwich Construction

Sandwich construction is a tricky business. Joints require tapering of the core to permit the skins to be brought together, local reinforcement is generally needed around connection points, and it is expensive. Nevertheless, the results seem to justify the problems.

The Lotus 79—a single-skinned aluminum monocoque—provided a torsional stiffness of 3,000ft-lb per degree from 95lb of structure. Changing to an aluminum skin/aluminum honeycomb sandwich saved 10lb and increased stiffness to 5,000ft-lb per degree. The all-composite 1980 chassis, using carbon and Kevlar with a Kevlar honeycomb core, achieved 10,000ft-lb per degree from just 75lb. On a stiffness-to-weight basis, the composite sandwich was more than twice as efficient as its metal precursor. More recent F1 chassis achieve well over 25,000ft-lb per degree.

On a stiffness-to-weight basis, the composite sandwich was more than twice as efficient as its metal precursor. More recent F1 chassis achieve well over 25,000ft-lb per degree.

Chapter 11

Unobtanium

Exotic Alloys

Beyond the range of alloy steels dealt with in chapter 4, there are many exotic alloys originally developed either as tool steels (offering a combination of exceptional strength and toughness at moderate temperatures), or else for internal parts of jet engines or for the petrochemical industry (to provide a combination of strength and resistance to "creep" and corrosion at high temperatures). These last are referred to as "super alloys." Some are based on iron, others on nickel or cobalt.

Without exception, these very expensive materials are produced and used in very small quantities. In race cars they may be used for exhaust valves, turbocharger turbines, and similar high-heat parts, and for very highly stressed components like rod bolts.

Of iron-based exotic alloys for moderately high temperature use, one of the more common is "H-11," containing 0.35 percent carbon, 5 percent chromium and 1.5 percent molybdenum. H-11 can be heat treated to a UTS of 290ksi. In slightly more ductile and damage-tolerant tempers it is used for aerospace bolts of 220 or 260ksi. Such fasteners are frequently used in race cars as connecting rod bolts. Aftermarket wrist pins for race engines made from H-11 are also available.

These strength levels are also achievable with conventional alloy steels, such as 4340, but the massive increase in the amount of the alloying elements—plus more stringent limits on impurities and other quality-control specifications—

reveals itself in the fatigue resistance of H-11 compared to 4340, heat-treated to the same nominal strength level of 220-250ksi. Under a cyclic stress varying from 11–115ksi, the H-11 material exhibits twenty times the life of 4340.

"Vasco Jet-MA," a proprietary alloy of Vanadium-Alloys Steel Co., can be heat treated to 350,000psi in sections up to at least 7/8in. In a slightly "softer" temper, suitable for a 300ksi aerospace bolt (the yield strength is at least 240ksi), this material still exhibits 7-8 percent elongation!

Despite optimistic predictions twenty years ago about steels of half a million psi or more, stubborn problems of stress corrosion cracking have turned attention toward materials with higher inherent corrosion resistance than those consisting mostly of iron. This problem as it affects certain high-strength steel alloys has become of serious concern with the advent of fasteners in the over-200,000psi range. It has necessitated, for instance, the wholesale scrapping of certain steel bolts in the landing gear of a widely used jumbo jet and the substitution of exotic, non-ferrous alloy replacements at enormous expense. Also, there are relatively few applications calling for such enormous strength levels that do not also combine a requirement for strength beyond the 500–800 degree F limit of these tool steels. The resulting "super alloys" are the types found in high performance exhaust valves, etc.

One material long used for jet engine parts is "A-286," which contains about 56 percent iron and

Despite optimistic predictions twenty years ago about steels of half a million psi or more, stubborn problems of stress corrosion cracking have turned attention toward materials with higher inherent corrosion resistance than those consisting mostly of iron.

so qualifies as steel, but with high temperature properties enhanced by 15 percent chromium, 25 percent nickel, plus titanium, aluminum, and other alloying elements, it may also be regarded as a super alloy. A-286 has room temperature ultimate strengths up to 130ksi after heat treatment, and subsequent work hardening can raise that figure to 200ksi. Useful strength (at least 63ksi UTS) is retained to 1200 degrees F.

Another jet alloy— "Inconel 718"—combines the mechanical strength of H-11 with excellent corrosion resistance and the temperature tolerance of A-286; it can typically provide 100+ksi at 1200 degrees for hundreds of hours. Though it contains 18 percent iron, Inconel 718 is 53 percent nickel, so is considered a nickel based alloy. It also includes 18 percent chromium, plus other elements, including significant traces of aluminum and titanium. (While the details of metallurgy are beyond the scope of this book, we should note that it is the combination of titanium and aluminum, typically in amounts of a couple of percent, that is responsible for the retention of strength at high temperatures shown by the nickel-based super alloy.)

Other super alloys consisting almost entirely of nickel and chromium, together with small amounts of aluminum and titanium, include other grades of Inconel, various "Nimonic" alloys, and "Waspalloy." They provide ultimate strengths up to 135ksi at room temperature, up to 100ksi at 1200 degrees F, and some are serviceable to 2000 degrees. Such alloys are used for exhaust valves, turbocharger turbines, critical parts of brake systems, etc., as well as springs and fasteners for high heat areas.

Though less strong at room and medium-high temperatures, certain cobalt-based alloys exhibit exceptional retention of hardness and strength at very high temperatures. These alloys are identified by a bewildering assortment of letter-and-number combinations like X40, HS151, J-1570, etc. (the English call all this stuff "Stellite"), but all are based on 60 percent or more of cobalt, plus nickel and chromium. Apart from use as specialized cutting tools, these "Stellite-like" alloys are frequently used as flame-sprayed facings for the contact faces, tips, and seats of valves. Some can endure for thousands of hours at temperatures above 1500 degrees F, albeit with strength levels down to a few ksi.

At less extreme temperatures, some cobalt/nickel/chromium alloys are being used for ultra-high-strength engine bolts. One of these, trade-marked "Multiphase MP35N" by SPS Technologies, Inc., was originally developed by DuPont for use in very caustic environments in the pulp and paper industry. MP35N (with typical room temperature properties of 286ksi UTS and 255ksi, with 11 percent elongation, a fatigue endurance limit of 140ksi, and complete immunity to stress corrosion cracking), has been used for rod bolts in Formula 1 racing. A newer alloy "MP159" (a registered trademark of SPS Technologies, Inc.), though perhaps not quite as strong at room temperatures, is serviceable to 1100 degrees F.

Aluminum-Lithium

A recent addition to the range of aluminum alloys is a group containing Lithium. ALCAN International has done much of the work in this field, in association with Britain's Royal Aeronautical Establishment, and has registered the trademark "LITAL" to describe these new products.

Unlike traditional alloying substances, which increase strength but have essentially no effect on the density or stiffness of the base metal, lithium significantly improves both. As a pure element, lithium has an astonishingly low density—just 0.019lb/cu-in, which sounds more like wood than metal. Just 2-3 percent of it added to aluminum, together with copper and magnesium in amounts comparable to some other aluminum alloys, reduces the density of the end product by 7-10 percent, while increasing stiffness by a roughly equal amount. The combined effect is an increase in specific stiffness of about 20 percent, which represents a breakthrough of sorts for metals.

LITAL "A" and "C" have strength properties in all loading modes comparable to 2024-T6 aluminum, while LITAL "B" basically matches 7075-T6. These alloys are already being used in a few non-critical applications (seat brackets, for instance) in commercial aircraft, as part of ongoing field tests. Needless to say, this material is more expensive than traditional aluminum alloys and the lithium component is toxic. More widespread application will depend, then, not just on economic considerations but also on satisfactory solutions to the health and environmental hazards presented by the lithium itself and by the processes involved in producing the alloy.

Magnesium-Lithium

The same concerns about the toxic hazards of lithium that inhibit the development of lithium-aluminum also apply to magnesium alloys that include lithium. One containing 14 percent lithium and 1.25 percent aluminum progressed beyond a purely experimental stage, being used as a spacecraft material. With a density of just 0.049lb/cu-in, Mg-14Li-1.25Al is perhaps the lightest metal ever used for a structural purpose, and that low density combines with an elastic modulus of 6.2 msi to yield a specific stiffness comparable to aluminum-lithium. The density considered together with a yield strength of 14.9ksi, however, results in a specific strength virtually identical to conventional wrought magnesium.

Beryllium-Aluminum

An instance of curious combinations of metals to achieve properties unobtainable from any of the constituents is a 62 percent beryllium/38 percent aluminum alloy developed by Lockheed. With a yield strength of 60ksi, an elastic modulus of 29msi and a density of 0.076lb/cu-in, the resulting "Lockalloy" has specific strength and stiffness lying between those of the two component metals.

Beryllium

With a density 0.067lb/cu-in, yet a Young's modulus of 42msi (50 percent higher than that of steel), beryllium has more than six times the specific stiffness of traditional structural metals. Indeed, its stiffness-to-weight ratio is greater than any other structural material having useful strength in multiple directions. (Uni-directional high-modulus carbon fiber laminates exceed beryllium's performance in this regard, but only along their "primary" axis—they cannot resist significant loads coming from other directions.)

Beryllium is also outstanding in terms of specific strength. With ultimate and yield strengths as much as 120ksi and 50ksi, respectively, yet with the same density as magnesium, beryllium is surpassed in specific strength only by high-strength titanium, and a few composites and super alloys. It also has excellent fatigue resistance, and retains its properties to very high temperatures—it is serviceable to 1550–1700 degrees F.

But these promising specifications are overwhelmed, for most applications, by three major factors. First, beryllium is tremendously expensive. Second, it is toxic—inhaled dust from machining operations, or even a sliver in the skin causes serious medical problems. Finally, beryllium has poor fracture toughness (it is brittle, in other words) and is extremely sensitive to surface flaws—in spacecraft structures made from the light metal it has been found necessary to use a "chemical milling" process to smooth the micro-

Interestingly, it was not beryllium's strength or stiffness that was sought in the brake rotors, but rather its astonishing thermal qualities.

scopic marks left by machining operations, in order to avoid cracks propagating from those marks.

In the aerospace field, where the value of weight saving is measured in thousands of dollars per pound, beryllium has been used for fasteners (a beryllium bolt will float in salt water!) and for various structural parts, including 146lb of it in some hinges on the space shuttle. One of the very few race car applications of beryllium (perhaps the only one) was for the brake rotors on a Porsche sports car entered in the European hill-climb championship in 1967. In this application the beryllium was plated with nickel, to prevent direct contact between the brake pads and the beryllium surface, and thus the risk of producing a toxic dust. Interestingly, it was not beryllium's strength or stiffness that was sought here, but rather its astonishing thermal qualities.

As explained in chapter 6, there are two factors that affect the selection of a material for brake rotors—its maximum service temperature and its "specific heat" (how much heat it takes to raise the temperature of 1lb of the stuff by 1 degree). On both scores, beryllium is outstanding. With a specific heat of 0.45 BTU per pound per degree F, 1lb of the metal soaks up about twice as much heat as a pound of aluminum for a 1 degree temperature rise, and about four times as much as a pound of iron. And, as mentioned, beryllium provides useful strength at temperatures above the melting point of other light metals. Thus, prior to its race car use by Porsche, beryllium had been selected for brake rotor material on various fighter jets and on the C5A "Galaxy" aircraft.

Carbon/Carbon

If the cost, toxicity, and other drawbacks of beryllium are so severe that the only race car ap-

plication it was ever likely to see was for brake rotors, then we're probably never going to see it again. Since that venture by Porsche in 1967 (see "Beryllium," above), a new material, carbon fiber-reinforced carbon, also called "carbon/carbon" has come onto the scene. Compared to beryllium, it offers even greater advantages over traditional brake rotor materials, and for the same reasons.

The specific heat of carbon/carbon ("C-C," for short) rises with temperature from about two-thirds less than beryllium at room temperature, to roughly equal to it at that metal's melting point. The strength of C-C, though, remains almost constant up to 4,500 degrees, at which point it "sublimes"—like dry-ice at room temperature, it does not melt but rather turns directly into gas.

Taking into account its density (the same as magnesium) and its specific heat, a pound of carbon/carbon is able to absorb more than five times as much heat as a pound of iron, seven times as much as aluminum, and 60–70 percent more than a pound of beryllium, when each material is at its working temperature limit. Accordingly, the same logic that led to beryllium replacing ferrous metal for the brakes on some aircraft (the Lockheed C5A, for instance), has now led to the replacement of beryllium by carbon/carbon. The newer C5B uses carbon/carbon brake rotors, as does the supersonic Concorde, the Boeing 747-400, and numerous other commercial and military aircraft. Though C-C brakes were recently banned from IMSA sports cars and World of Outlaws sprint cars on the grounds of cost, they are now universal in Formula 1. Clutches using the same material are common in many racing classes.

While cheaper than beryllium, C-C is still terribly expensive—well over $100 per pound, largely due to the method of production. Two different methods are used, but each begins with a "preform" of carbon fiber, consisting either of individual layers of cloth stacked together or produced as a unit by three-dimensional weaving. The preform may be saturated with a plastic resin (perhaps phenolic), then put through a lengthy series of baking cycles at high temperatures. More resin is added at various stages, as the resin components other than carbon are driven off, until all that remains is pure carbon. Alternatively, the preform remains "dry," and the oven filled with a carbon-rich gas such as methane. This vapor-deposits carbon onto the surface of each fiber, until

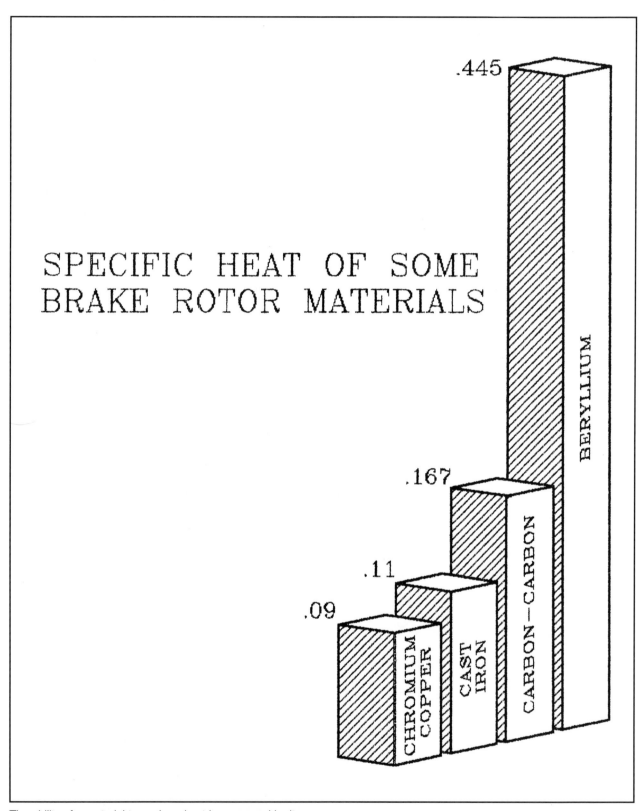

SPECIFIC HEAT OF SOME BRAKE ROTOR MATERIALS

.09 — CHROMIUM COPPER
.11 — CAST IRON
.167 — CARBON—CARBON
.445 — BERYLLIUM

The ability of a material to soak up heat is expressed by its "specific heat," a property that is obviously significant in selection of a brake rotor material. On this basis, beryllium far outstrips other candidate materials.

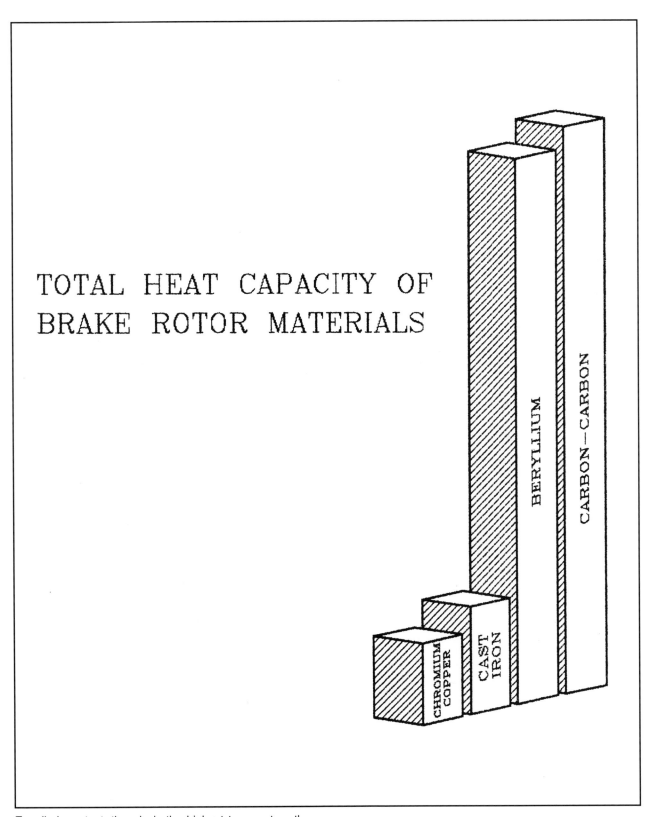

TOTAL HEAT CAPACITY OF BRAKE ROTOR MATERIALS

CHROMIUM COPPER

CAST IRON

BERYLLIUM

CARBON-CARBON

Equally important, though, is the highest temperature the material will tolerate. When that is taken into account, the reason top rank professional teams use carbon/carbon brake rotors becomes obvious.

all the empty spaces are filled.

Though increased production may lower the cost of C-C, the time and energy needed to produce it will probably ensure that it remains expensive. Nevertheless, some of the high initial cost can be justified by its durability—carbon/carbon racing clutches have lasted up to sixteen times longer than more conventional ones. And for brakes there is simply no alterna-tive material which offers the combination of heat tolerance, weight saving, and stable friction characteristics. (C-C rubbing on itself has a friction coefficient that rises sharply from cold, but around 200 degrees F and above it stabilizes at about 0.35–0.45.) Even at a rumored $1,000 per corner, if C-C brakes reduce total weight by 30lb, as claimed, the cost for each pound saved seems reasonable.

Brake rotors of carbon fiber-reinforced carbon (carbon/carbon, for short) soak up more heat per pound than cast iron, can operate at twice the temperature of iron, yet weigh the same as magnesium. *Doug Gore, courtesy* Open Wheel *Magazine*

Ceramics

A great many commonly occurring minerals can be used to make hard, shaped goods by grinding the minerals to a powder, then compacting them into the desired shape followed by firing in an oven, to "sinter" the particles together. The minerals are often compounds of silicon or a metal, together with oxygen, carbon, or nitrogen—aluminum oxide, silicon carbide, etc. To facilitate the forming process, water, waxes, or resins may be used to fluidize the powdered raw material, and traces of these binders, as well as internal voids, may also be found in the finished product.

Depending on the pressures and temperatures involved in the compacting and firing processes, the result may consist mainly of a glassy matrix surrounding crystals of the same material, or may be composed almost entirely of crystals, atomically bonded together. The mechanical properties of the resulting ceramic depend strongly on the specific raw materials chosen, on the details of the sintering process, and on the size and number of internal voids and inclusions. Results can vary from ordinary tableware to high-strength engineering ceramics.

Advanced engineering ceramics are effective heat insulators, have excellent friction characteristics, combine the strength and density of aluminum with stiffness superior to that of steel, and maintain those properties at incandescent temperatures. This sounds like great stuff for engine parts! Indeed, ceramic coatings and inserts have been widely used in diesels, at least a million ceramic turbocharger turbines have been sold, and there's at least one ceramic part in every gasoline engine—the spark plug insulator.

There's a snag, though—ceramics are brittle, having neither the ductility of metals nor any other toughening mechanism, comparable to the effect of fibers in composites. That isn't so bad for a part like a plug insulator or a piston crown insert, where the shape is simple and the forces are mostly compressive, or even a turbo wheel which experiences large tension loads, but is spared mechanical shocks. But many internal engine parts are stressed in tension, have rapid changes in contour, and experience violent impacts; under these conditions, ceramics may fail catastrophically. Thus, most early attempts to exploit the benefits of ceramics in engines resulted in loud bangs and clouds of shrapnel.

Nevertheless, the attractions of ceramics inspired much development work directed aimed at fortifying their unreliable tension properties. One area of focus has been improving the uniformity of particle size and the purity of the raw materials. Another is new high-pressure sintering techniques to reduce or eliminate internal flaws. This work has yielded ceramic materials suitable for use in race engines.

Advanced engineering ceramics are effective heat insulators, have excellent friction characteristics, combine the strength and density of aluminum with stiffness superior to that of steel, and maintain those properties at incandescent temperatures.

For the ceramic coatings (usually plasma-arc sprayed) used in diesel engines, the requirements are for a material that insulates well in thin sections and that matches the expansion of the underlying metal. Zirconia (an oxide of zirconium metal) is widely used for this purpose, and meets these requirements fairly well, combining thermal expansion about half that of aluminum with barely 1 percent of that metal's conductivity. For "monolithic" (one-piece) parts, however, zirconia is less attractive than some other ceramics.

In view of their brittle nature, larger factors of safety have to be employed with ceramics than with tougher materials. After taking fatigue into account, it may be reasonable to design a steel or aluminum part to be working at, say, one-third of its UTS. For ceramics, the safe working stress is a much smaller fraction of the typical UTS—more like 6–15 percent. When this new strength figure is compared to its density, zirconia does poorly compared to metals. Silicon carbide and silicon nitride, which roughly match aluminum and steel

Robert Larsen, an employee of Argonne National Laboratory, inspects silicon nitride exhaust valves for a stock car race engine. Despite the theoretical promise of monolithic ceramic components, their brittleness means that when they fail, they do so catastrophically. *Argonne National Laboratory*

by these measures, offer more promise. Pushrod tips, valve lifters, end even complete valves have been made of these materials. In one racing engine, lightweight silicon nitride exhaust valves allowed an increase in revs from 10000–12500 rpm before valve float set in.

Most sensational of all is the use of ceramic pistons in Formula 1 race engines, specifically the Honda engines used in 1991 by the McLaren team. Although the material used (probably silicon nitride) is slightly more dense than aluminum, the ceramic's greater strength and higher Young's modulus, especially at elevated temperatures, permitted use of very thin sections, according to the material supplier. The resulting reduction in piston weight was the biggest advantage claimed for the use of ceramics in this application, permitting an 800–900rpm gain. While the whole

issue remains cloaked in secrecy, it seems that the main reason the scheme was abandoned was because of excessive oil consumption.

A laboratory test engine with ceramic pistons and cylinders had been successfully run in 1983. In view of the very slight thermal expansion of the material used (pistons and cylinders both of silicon carbide in some tests, silicon nitride pistons in others), it was possible to run the engine with very tight piston clearance—just 0.002in, cold, closing to 0.001in when hot. That, in turn, permitted running the engine without piston rings, which significantly reduced friction.

This is a double-edged sword, however—with ceramic pistons, ringless operation is not only possible, it may be necessary. For one thing, the ring grooves create zones of high stress concentration, which must greatly reduce the reliability

While monolithic ceramics and cermets have proved troublesome in performance engines, thin coatings of such materials have overcome early teething problems and are now an accepted technology in many forms of racing. These pistons and valves have had their heat-exposed surfaces coated with a zirconia/magnesium-zirconate/refractory metal cermet; piston skirts are also treated with a low friction coating. *Swain Technology*

of the ceramic piston. Also, at the temperatures experienced by ceramic pistons, lubricating oils would carbonize, sticking the rings in their grooves. It appears that when Honda employed ceramic pistons they too ran them ringless—certainly the engine was a conspicuous "smoker," and when the organizers banned the supplementary oil supply carried by the McLarens, a completely new engine design was introduced.

Ceramic Matrix Composites and Cermets

Various attempts have been made to reduce the brittleness of ceramics. One technique involves mixing powdered metals into the raw materials. The resulting "cermets" may be mostly metal or mostly ceramic, or anywhere in between. Their properties, quite reasonably, also fall between those of metals and of ceramics, depending on the constituents and proportions of the mix. Tungsten carbide cutting tools are cermets—apart from the ceramic compound formed from tungsten and carbon (WC), they also contain a small proportion of columbium metal. "Cerametallic" brake linings are another example, one in which advantage was taken of an interesting feature of cermets—the possibility of brazing the material to a metal backing. During the early nineties, an engine valve comprised of 90 percent titanium (metal) and 10 percent titanium dioxide (ceramic) was tested for use in stock car racing, in the hope of improving on the marginal durability of titanium when used for exhaust valves in this application. Extensive tests confirmed the superior fatigue and impact resistance of the cermet valve at room temperature, and its 35 percent greater high temperature strength than titanium metal when loaded statically. Combining both thermal and impact loads in a running engine, however, yielded only expensive noises, much to the developers' dismay.

Greater success has been achieved with thin cermet coatings, in fact most "ceramic" engine coatings are actually cermets. Zirconia and alumina (Aluminum oxide) are common ceramic components; suppliers are reluctant to identify the metallic phase, though aluminum and refractory (heat resistant) alloys like cobalt, nickel, and chromium are most likely. Such coatings are routinely applied to piston crowns, valve faces, and combustion chambers in racing engines, yielding slight gains in power and fuel consumption, and substantial improvements in the life of these highly stressed components.

More recent efforts to improve the toughness of ceramics have taken advantage of the crack stopping mechanism of fibers within a matrix (see chapter 9—"Composites, Part 2"). The resulting materials are "ceramic matrix composites," or CMCs. One method for producing CMCs, called "directed metal oxidation," begins with a preform of ceramic fibers, say aluminum oxide, either woven into a three-dimensional piece or laid up from multiple plies of two-dimensional cloth. The preform, set on top of a quantity of metal (typically aluminum), is placed inside a gas-permeable mold, and the assembly is then heated in a controlled atmosphere furnace. The metal flows through the preform by capillary action, and reacts with the furnace atmosphere to produce a ceramic—aluminum in an oxygen atmosphere, for example, would produce aluminum oxide. Thus, the matrix is effectively "grown" around the reinforcement.

As an alternative to continuous fibers, the reinforcement can take the form of thin flakes, called "platelets." A silicon carbide-reinforced aluminum oxide composite produced in this way has shown a room temperature flexural (bending) strength of 63ksi, while at 2200 degrees F the flexural strength remained 55ksi. Cam followers, valve seats and complete camshafts made from CMCs of different compositions have been tested in engines with encouraging results.

Metal Matrix Composites

Adding reinforcement to plastic resins or ceramics provides a toughening mechanism to overcome their inherent brittleness, but metals already possess toughness, by virtue of their ductility. When metals are used as the matrix for a composite material—producing metal matrix composites (MMCs)—the objective is usually to increase such properties of the matrix metal as strength, stiffness, and temperature resistance.

To date, most trials of MMCs in the automotive field have been based on preforms as described above (see "Ceramic Matrix Composites"), using aluminum as the matrix. Magnesium matrix MMCs have also been tried. Using pressure casting at moderately high pressures, the room temperature ultimate and yield strengths of 334 alloy aluminum were raised 13 percent and 32 percent, respectively, with the addition of 18 percent of an "aluminosilicate" ceramic fiber—48 percent aluminum oxide, 52 percent silicon dioxide. At higher temperatures, the gain was more significant—65 percent and 90 percent, respec-

Some Ceramics & MMC's Compared to Metals

Material	Safe Stress ksi	E msi	D lb/cu in	Conductivity (relative)	Expansion (ppm/degree F)
Steel	30	30	0.30	32	7
Aluminum	10	10	0.10	100	12
Silicon Nitride	8–19	42	0.12	3	1.7
Silicon Carbide	7–11	59	0.11	3.4–4	1.9–2.2
Aluminum Oxide	7–8	52	0.12	8	4.4
Aluminum/20% SiC		14	0.10	–	–
Aluminum/60% SiC		30	0.11		–4.7–6
Aluminum/50% aluminum Oxide		23	0.11	–	7.5

tively, at 536 degrees F. Similar work with AZ91 magnesium, using 20 percent aluminum oxide fiber reinforcement, showed an increase in UTS of better than 50 percent at room temperature, though the strengthening effect did not increase as much at higher temperatures as in the case of aluminum.

While pressure casting makes rapid production possible, and permits the casting of complex shapes, the process limits the length of fibers that can be used in the preform—long fibers get crushed into smaller ones, due to the pressure. A new proprietary process, involving special atmospheres and fluxes, allows the filling of the preform with molten metal at reasonable rates, without the use of pressure. That, in turn, permits the use of long-fiber preforms, and much higher fiber fractions.

Using these processes to produce aluminum containing 60 percent silicon carbide yielded strength levels and density about equal to un-reinforced aluminum, but stiffness equal to steel. With 50 percent aluminum oxide in a different aluminum alloy, the stiffness declined to about 22msi (about double that of aluminum, two-thirds that of steel), but strength increased to 65ksi. A valve retainer made from this material has been tested for racing use. A fuel pump pushrod made from 60 percent silicon carbide fiber-reinforced aluminum is widely available.

Apart from the mechanical strength and stiff-ness provided by the addition of ceramic fibers, a major advantage of MMCs is the ability to modify the thermal expansion and wear characteristics of the finished product. The aluminum cylinder block of the Honda Prelude, for example, has a belt of reinforcement around the top of each cylinder bore, consisting of a mixture of carbon fiber and ceramic fiber. This local region of MMC exhibits wear characteristics comparable to a cast iron bore, yet is compatible in thermal expansion with both the piston material and the remainder of the aluminum block.

Preforms permit medium to high fiber fractions, but they imply parts produced to finished or near finished shape and size. To produce an MMC in the form of a raw material in bulk that can be cast to shape later, the fibers must be short, and small in number. ALCAN has registered the trademark "Duralcan" to describe such a general purpose MMC. While they are secretive about the method of production (it has been described as "like throwing grapes in Jello"), they are very optimistic about the prospects for automotive applications. "Duralcan" containing 10 percent silicon carbide has room temperature properties only a little better than conventional grades of aluminum, but retains its strength sufficiently well at elevated temperatures that it is being touted as a possible replacement for cast iron in brake rotors. "Duralcan" costs four to seven times as much as aluminum.

Appendix

Selected Bibliography/
Suggestions For Further Reading

Adams, A.A. "Composite structures for automobiles—Lotus experience," in *ASM Advanced Composites Conference Proceedings, December 2-4 1985.*

Agranoff, Jeff (ed). *Modern Plastics Encyclopedia.* McGraw-Hill. N.Y., 1976.

ALCAN International. "Current Data for Aluminium-Lithium Alloys" [pamphlet]. Gerrard's Cross, Buckinghamshire.

Allen, P.G. *Technical and commercial considerations for the application of titanium in automotive engines,* in: *Proceedings of the Institution of Mechanical Engineers.* v. 208 part D.

Aluminum Association, Inc. *Aluminum Standards and Data 1988.* The Aluminum Association, Inc. Washington, 1988.

Aluminum Co. of Canada Ltd. *Handbook of Aluminum,* 2nd edition. Aluminum Co. of Canada Ltd. Montreal, 1961.

Atlas Steels Ltd. *Technical Data.* Atlas Steels Ltd. Welland, Ontario.

Aviation Week and Space Technology. "Improved materials in new landing gear lighter and stronger," in: *Aviation Week and Space Technology,* December 12, 1988.

Beckman, Hans-Dieter & Hermann Oetting. Fiber reinforced plastics for lightweight engine parts. SAE 850520.

Benjamin, B.S. *Structural Design With Plastics.* Van Nostrand Reinhold. NY, 1969.

Bert, Charles W. & Karl H. Bergey. Structural cost effectiveness of composites. SAE 730338.

Bingelis, Tony. "Aircraft plywood... how to use it," in: *EAA Aircraft Building Techniques—Wood.* EAA Aviation Foundation. Oshkosh WI, 1991.

Bolster, John. *French Vintage Cars.* Autosport. London, 1964.

Broverman, I. *Evaluation of cast chromium copper for automotive disk brake rotors.* IITRI-B9634-2 summary report. IIT Research Institute.

Broverman, I. *Evaluation of a beryllium copper alloy for automotive disk brake rotors.* IITRI-B9634-1 summary report. IIT Research Institute.

Bruno, P.S. & D.O. Keith & A.A. Vicario Jr. "Automatically woven three-dimensional composite structures," in: *SAMPE Quarterly,* v. 17, no. 4, July 1986.

Case, Richard K. & Darrel D. Lemon & David W. Paule. "Beryllium: the lightweight contender," in: *Machine Design,* June 7, 1984.

Cedar, Dennis A. Fabrication and assembly of a graphite fiber reinforced plastic vehicle. SAE 790029.

Chi, F.K. et al. Short ceramic fiber reinforced aluminum alloy. SAE 870440.

Clauser, H.R. (ed). *Encyclopedia Handbook of Materials, Parts and Finishes.* Technomic. Westport, CT., 1976.

Cox, H.L. *The Design of Structures of Least Weight.* Pergamon Press. Oxford.

Crane, Dale. *Aircraft Corrosion Control.* Aviation Maintenance Publishers. Basin, WY, 1979.

Crane, F.A.A. & J.A. Charles. *Selection and Use of Engineering Materials.* Butterworths. London, 1989.

Cross, S.L. Graphite composite materials applications in aircraft structures. SAE 750533.

Davis, James. The potential for vehicle weight reduction using Magnesium. SAE 910551.

Delmonte, John. *Technology of Carbon and Graphite Fiber Composites*. Van Nostrand Reinhold.

Department of Defense (USA). *MIL-HDBK-5D—Metallic Materials for Aerospace Vehicle Structures*. Washington, 1983.

Dietz, Albert G.H. (ed). *Composite Engineering Laminates*. MIT Press. Cambridge, MA, 1969.

Dinwoodie, J. Automotive applications for MMCs based on short staple alumina fibres. SAE 870437.

Du Pont Co., *Technical Symposium V*. Du Pont de Nemours, Inc., Industrial Fibers Divisionn.

Dwivedi, R. et al. Applications of metal matrix composites in high performance car engines. SAE 911770.

Ederer, Ulf G. Design criteria and experience with crankshaft bearings in medium speed engines. SAE 851196

Engel, Ulrich. Development and testing of new multi-layer materials for modern engine bearings: Part 1—copper-tin bonding and intermediate layers. SAE 860354

Engel, Ulrich. Development and testing of new multi-layer materials for modern engine bearings: Part 2—copper-lead three layer bearings with sputtered overlay. SAE 860648

Flynn, G. & J.W. MacBeth. A low friction, unlubricated, uncooled ceramic diesel engine—Chapter 2. SAE 860448.

French, C.C.J. Ceramics in reciprocating internal combustion engines. SAE 841135.

Frere, Paul. *The Racing Porsches*. Arco. New York, 1973.

Rules for Surveying and Testing of Plywood for Aircraft. Germanischer Lloyd. Berlin, 1953.

Gibson, D.W. & G.J. Taccini. Carbon/carbon friction materials for dry and wet brake and clutch applications. SAE 890950.

Gill, R.M. *Carbon Fibres in Composite Materials*. Iliffe. London.

Gilson, Robert D. Beryllium brake experience on the C-5A airplane. SAE 710427.

Goddard, David M. & Walter R. Whitman & Robert L. Pumphrey. Graphite/magnesium composites for advanced lightweight rotary engines. SAE 860564.

Gordon, J.E. *The New Science of Strong Materials*. Pelican. Harmondsworth, Middlesex, 1974.

Gordon, J.E. *Structures*. Pelican. Harmondsworth, Middlesex, 1978.

Gore, Douglas. "Carbon/carbon brakes," in: *Open Wheel*, December, 1990.

Gowen, Edward F. & Ronald Waeltz. *Report No. 659—A 300,000psi Minimum Tensile Strength Fastener Assembly For Use to 550 Degrees F.* SPS Technologies. Jenkintown, PA, 1961.

Gowen, Edward F. & Richard A. Walker. *Report No. 730—Evaluation of 200,000psi Titanium Bolts and Locknuts HI TI 20 Series—FN T20 Series*. SPS Technologies. Jenkintown, PA, 1963.

Guldberg, S. & H. Westengen & D.L. Albright. Properties of squeeze cast, Magnesium-based composites. SAE 910830.

Hagan, Frank C. and Marshall R. Mazer. *Report No. 1787—The Evaluation of a 260,000psi Tension Fastener System Made from Multiphase Alloy MP35N*. SPS Technologies. Jenkintown, PA, 1968.

Hagiwara, Yoshitoshi & Kyo Takahashi. Development of surface treatment and application to mass-production of titanium connecting rods. SAE 891769.

Hall, Stan. "On the use of Douglas fir as a substitute for spruce," in: *EAA Aircraft Building Techniques—Wood*. EAA Aviation Foundation. Oshkosh WI, 1991.

Hawley, Arthur V. Ten years of flight experience with DC-10 composite rudders—a backward glance. SAE 861674.

Hay, N. et al. "Design study for a low heat loss version of the Dover engine," in: *Proceedings of the Institution of Mechanical Engineers*. v. 200 part D, no. 1.

Hayashi, Tadayoshi & Hideaki Ushio & Masuo Ebisawa. The properties of hybrid fiber reinforced metal and its application for engine block. SAE 890557.

Henry, Alan. *Grand Prix Car Design & Technology in the 1980's*. Hazleton. Richmond, Surrey, 1988.

Hillesland, H. "Aerospace materials and process utilized in a race car," in: *Advanced Composites—Conference Proceedings December 2-4, 1985*. American Society for Metals.

Hodges, David. *The Ford GT40 Prototype and Sports Cars*. ARCO Publishing, New York, 1970.

Huntington, Roger. *Design and Development of the Indy Car*. HP Books, Tucson, AZ, 1981.

Ingham, H.S. Jr. *Flame-sprayed coatings*, in:

Dietz, Albert G.H. (ed) *Composite Engineering Laminates.* MIT Press, Cambridge MA, 1969.

Kent, M.J. "Titanium," in: *Design Engineering,* April 1964.

Kiefer, W.R. et al. Fabrication of automotive body components in GrFRP. SAE 790028.

Kissane, M.L. High performance brakes for CERV-III. SAE 910576.

Kojima, T., & N. Kato & N. Ishida & M. Taniguchi. "Ceramic applications and related technology," in: *Proceedings of the Institution of Mechanical Engineers.* v. 204 part D, 1990.

Kuch, Ingo & Gudrun Delonge-Immik. Development of FRP rear axle components. SAE 910888.

Kurrein, Max. *Plasticity of Metals.* Charles Griffin & Co. New York, 1964.

Langston, Paul. Application of advanced composites in civil aircraft. SAE 861673.

Ludvigsen, Karl. *Gurney's Eagles.* Motorbooks International. Osceola, Wisconsin, 1992.

Marsh, Cedric. *Strength of Aluminum,* 5th edition. Aluminum Co. of Canada Ltd. Montreal, 1983.

Marshall, Andrew. "Composite Basics," (a series of articles appearing in *Homebuilt Aircraft* between February, 1983 and August, 1984).

Marshall, Andrew. *Practical Sandwich Structures/Advanced Composites.* Marshall Consulting, Inc. Walnut Creek, CA, 1982.

Megson, T.H.G. *Aircraft Structures for Engineering Students.* Edward Arnold. London, 1972.

Mezger, Hans. Engineering the performance car. SAE 700678.

Mezger, Hans. "The development of the Porsche type 917 car," in: *Proceedings of the Institution of Mechanical Engineers.* v. 186, 1972.

Mezoff, John G. Magnesium for automobiles, in perspective. SAE 800417.

Miska, Kurt H. High performance cars demand high performance materials. SAE 780423.

Murphy, A.J. "Materials in aircraft structures," in: *Journal of the Royal Aeronautical Society.* v. 70, January 1964.

Niles, Alfred S. & Joseph s. Newell. *Airplane Structures.* Wiley. NY, 1943.

Noakes, Keith. *Build to Win.* Osprey. London, 1988.

O'Rourke, B.P. "The use of composite materials in the design and manufacture of Formula 1 racing cars," in: *Proceedings of the Institution of Mechanical Engineers.* v. 204 part D.

Payne, R.J. (ed). *Plywood Construction Manual.* Council of the Forest Industries of British Columbia. Vancouver BC, 1969.

Peery, David J. *Aircraft Structures.* McGraw-Hill. New York, 1950.

Pratt, G.C. & W.J. Whitney. Progress with aluminum-lead crankshaft bearing alloys. SAE 890552

Pratt, G.C. & C.A. Perkins. Aluminium based crankshaft bearings for the high speed diesel engine. SAE 810199

Rainbolt, Jack D. Effects of disk material selection on disk brake rotor configuration. SAE 750733.

Rhee, S.K. & R.M. Rusnak & W.M. Spurgeon. A comparative study of four alloys for automotive brake drums. SAE 690443.

Rhee, S.K. & R.T. DuCharme & W.M. Spurgeon. Characterization of cast iron friction surfaces. SAE 720056.

Rhee, S.K. & J.E. Byers. A comparative study by vehicle testing of copper alloy and gray iron brake discs. SAE 720930.

Roach, Thomas A. "Aerospace high performance fasteners resist stress corrosion cracking," in: *Materials Performance.* v.23 no.9, September 1984.

Roark, Raymond J. and Warren C. Young. *Formulas for Stress and Strain.* McGraw-Hill. New York, 1975.

Schiroky, Gerhard H. & Boyd W. Sorensen. "Structural ceramics you can count on," in: *Machine Design,* January 10, 1991.

Scott, Paul W. The cost effectiveness of weight reduction by advanced material substitution. SAE 861850.

Schmidt, Jack. *Report No. 176—EWB External Wrenching Bolt and FN 26 Featherweight Locknut Ultimate Tensile Strength 260,000psi—Minimum.* SPS Technologies. Jenkintown, PA, 1959.

Schumacher, Clifford A. Reinforced composites for piston engine components. SAE 892495.

Schwartz, R.T. and D.V. Rosato. *Structural-sandwich construction,* in: Dietz, Albert G.H. (ed) *Composite Engineering Laminates.* MIT Press, Cambridge MA, 1969.

Shackleford, James F. *The CRC Materials Science and Engineering Handbook.* CRC Press. Boca Raton.

Simons, Eric N. *An Outline of Metallurgy.* Hart. New York, 1968.

Smith, Carroll. *Engineer to Win.* Motorbooks International. Osceola, WI, 1984.

Smith, Clarence R. *Tips on Fatigue.* Experimental

Aircraft Association. Hales Corners, WI, 1974.

Soderquist, Joseph R. Certification of civil composite aircraft structure. SAE 811061.

Soderquist, Joseph R. Design/certification considerations in civil composite aircraft structure. SAE 871846.

SPS Technologies. *MP159 Bolting Alloy.* SPS Technologies. Jenkintown, PA.

SPS Technologies, Aerospace and Industrial Products Division. *Product Engineering Report No. 4718—Fastener Seminar 1974.* SPS Technologies. Jenkintown, PA. 1980.

Taylor, Edward. *Report No. 5817—MULTIPHASE Alloy Environmental Resistance.* SPS Technologies. Jenkintown, PA, 1980.

Thompson, D'Arcy Wentworth. *On Growth and Form.* Cambridge U. Press. London. 1961

Timoney, S. & G. Flynn. A low friction, unlubricated SiC diesel engine. SAE 830313.

Van Valkenburgh, Paul. *Race Car Engineering and Mechanics.* Paul Van Valkenburgh. Seal Beach, CA. 1986.

Varrese, Francis R. *Report No. 5533.* SPS Technologies. Jenkintown, PA, 1974.

Yamamoto, Koichi et al. Development of three layers copper-lead bearings for higher speed automotive engines. SAE 910161

Zimmerman, O.T., & Irvine Lavine. *Handbook of Material Trade Names.* Industrial Research Service. Dover, NH. 1953.

Index